SCIENTIFIC WRITING IN A SECOND LANGUAGE

SECOND LANGUAGE WRITING
Series Editor: Paul Kei Matsuda

Second language writing emerged in the late twentieth century as an interdisciplinary field of inquiry, and an increasing number of researchers from various related fields—including applied linguistics, communication, composition studies, and education—have come to identify themselves as second language writing specialists. The Second Language Writing series aims to facilitate the advancement of knowledge in the field of second language writing by publishing scholarly and research-based monographs and edited collections that provide significant new insights into central topics and issues in the field.

BOOKS IN THE SERIES

The Politics of Second Language Writing: In Search of the Promised Land, edited by Paul Kei Matsuda, Christina Ortmeier-Hooper, and Xiaoye You (2006)

Building Genre Knowledge by Christine M. Tardy (2009)

Practicing Theory in Second Language Writing, edited by Tony Silva and Paul Kei Matsuda (2010)

Foreign Language Writing Instruction: Principles and Practices, edited by Tony Cimasko and Melinda Reichelt (2011)

Scientific Writing in a Second Language by David Ian Hanauer and Karen Englander (2013)

SCIENTIFIC WRITING IN A SECOND LANGUAGE

David Ian Hanauer and Karen Englander

Parlor Press
Anderson, South Carolina
www.parlorpress.com

Parlor Press LLC, Anderson, South Carolina, USA

© 2013 by Parlor Press
All rights reserved.
Printed in the United States of America

S A N: 2 5 4 - 8 8 7 9

Cataloging-in-Publication Data on File with
the Library of Congress and Parlor Press

1 2 3 4 5

ISBN Information
978-1-60235-379-4 (paperback)
978-1-60235-380-0 (hardcover)
978-1-60235-381-7 (Adobe ebook)
978-1-60235-382-4 (ePub)

Second Language Writing
Series Editor: Paul Kei Matsuda

Cover design by Paul Kei Matsuda and David Blakesley
Printed on acid-free paper.

Parlor Press, LLC is an independent publisher of scholarly and trade titles in print and multimedia formats. This book is available in paper, cloth and eBook formats from Parlor Press on the World Wide Web at http://www.parlorpress.com or through online and brick-and-mortar bookstores. For submission information or to find out about Parlor Press publications, write to Parlor Press, 3015 Brackenberry Drive, Anderson, South Carolina, 29621, or email editor@parlorpress.com.

Contents

List of Figures	ix
List of Tables	x
Acknowledgments	xi

1 The Dominance of the English Research Article in the Scientific World	**3**
1.1 English Literacy and the Generation of Scientific Knowledge	3
1.2 English and the Scientific Research Article	6
1.3 The International World of Science	9
1.4 The Critical-Pragmatic Approach to Second Language Science Writing	11
1.5 The Design of This Book	13
2 The Genre of the Scientific Research Article	**14**
2.1 A Genre Approach	14
2.2 Social Functions: Situating the Scientific Research Article	14
2.3 Producing a Research Article	19
2.4 The Structural Features of a Scientific Research Article	21
2.4.1 The Succinct Expression of Informational Complexity	22
2.4.2 Depersonalization through the Foregrounding of Knowledge Statements	23
2.4.3 Facilitating Navigational Clarity	25
2.4.4 Persuasion and Stance in the Research Article	29
2.5 The Research Article	30
3 Second Language Writing and the Research Article	**32**
3.1 Introduction	32
3.2 Socio-Economic and Geographic Context	33
3.3 Peripheral Participation	36
3.4 Interaction with the Journal Editors	40
3.5 Difficulties and Strategies of Writing	46
3.6 Contrastive Approaches and Language Community Characteristics	50
3.7 Summary	53

4 Methodology: Researching Spanish Speaking Scientists in Mexico — 55
   4.1 Introduction — 55
   4.2 Science in Mexico — 56
   4.3 Site Description — 58
   4.4 Participant Population — 59
   4.5 Method — 61
      4.5.1 Quantifying and Specifying the Difficulties of Writing a Research Article for Publication in a Second Language: Method and Analytical Approach — 61
      4.5.2 Developing Scientific Writing Expertise: Method and Analytical Approach — 63
          Question 1: Educational (Scientific Literacy) Background — 64
          Question 2: Personal Perceptions of Strengths and Difficulties — 64
          Question 3: L2 Scientific Writing Process — 65
          Question 4: Perceptions of Successful and Failed Writing Processes — 65
          Question 5: Perceptions of Helpful Solutions — 65
          Question 7: Open Discussion — 65
   4.6 Summary — 66

5 The Quantification and Specification of the Difficulties of Writing a Research Article for Publication in a Second Language: Survey Report — 67
   5.1 Introduction — 67
   5.2 Research Questions — 68
   5.3 Results — 68
   5.4 The Quantification and Specification of the Difficulties of Writing a Research Article in a Second Language: A Summary — 80
   5.5 Final Comments — 82

6 Developing Scientific Writing Expertise: Qualitative Individual Data — 83
   6.1 Introduction — 83
   6.2 Individual Data: Scientists' Narratives — 84
      6.2.1 Senior Researchers at the Research Institute — 84
      6.2.2 Senior Researchers at the University — 89
      6.2.3 Junior Researchers at the Research Institute — 97
      6.2.4 Junior Researchers at the University — 103
   6.3 Summary of Findings — 109
   6.4 Final Comments — 110

7 Developing Scientific Writing Expertise: Qualitative Group Data — 111
   7.1 Introduction — 111

7.2 Group Data: Trajectory of Learning Events 112
   7.2.1 Educational Trajectory: Bachelor degree 113
   7.2.2 Educational Trajectory: Masters Level 115
   7.2.3 Educational Trajectory: Doctorate Level 117
   7.2.4 Educational Trajectory: Professional Scientists 119
7.3 Trajectory of Learning Events within the Subgroups 121
7.4 Pedagogical Understandings: Participant-Scientists 126
7.5 Affective Responses 129
7.6 Summary of Qualitative Results 131

8 Facilitating Improved Scientific Writing in English as a Second Language 134
8.1 Aims and Underpinning Positions 134
8.2 Principles and Recommendations 136
   8.2.1 Long Term Commitment to Writing Education 136
   8.2.2 Differential Needs and Diversified Educational Interventions 137
   8.2.3 Multilayered Understanding of the Research Article 137
   8.2.4 Provision of Expert Support for Science and Writing 138
   8.2.5 Personalized, Continual, and Immediate Support for Research Article Writing 138
   8.2.6 Demystification of the Structures and Processes of Scientific Publication 138
   8.2.7 Broad Administrative, Institutional, and Financial Support 139
8.3 Educational Interventions 139
   8.3.1 Explicit Teaching 140
   8.3.2 Collaborative, Face-To-Face Interaction with Associates 144
   8.3.3 Expert and Peer Collaborations 147
      8.3.3.1 Mentoring Program 147
      8.3.3.2 Faculty Writing Circle 149
   8.3.4 Translation and Editing Services 154
8.4 Educational Program 156

9 Practical and Policy Implications of Supporting Second Language Scientific Writing 160
9.1 Introduction 160
9.2 The Challenges of Second Language Science Writing in English 161
9.3 The Role of Intervention 164
9.4 Principles of Support 166
9.5 Interventions by Different Stakeholders 167
   9.5.1 Universities 167
   9.5.2 National Higher Education Policy Makers 169

  9.5.3 Scientific and University Institutions   *170*
  9.5.4 Professional Scientific Bodies and Associations   *173*
 9.6 Policy Implications for Science   *173*

Notes   *175*
References   *177*
Index   *189*
About the Authors   *195*

List of Figures

Figure 4.1 Number of scientific articles published by Mexican scientists each year from 1982 to 2006. Source: CONACYT, 2009.   *57*

Figure 4.2 Percentage of countries' share of published articles in 2007. Source: CONACYT, 2009.   *57*

Figure 5.1 Means (figs. 5.1a & 5.1b) and standard deviations (fig. 5.1a) for self perception ratings on questions of difficulty, satisfaction, and anxiety in first language (Spanish) and second language (English) scientific writing (n=148).   *69*

Figure 5.2 Means (figs. 5.2a & 5.2b) and standard deviations (fig. 5.2a) for self perception ratings on questions of difficulty, satisfaction, and anxiety in first language (Spanish) and second language (English) scientific writing for faculty at a teaching university (TU) and a research institute (RI).   *71*

Figure 5.3 Means (figs. 5.3a & 5.3b) and standard deviations (fig. 5.3a) for self perception ratings on questions of difficulty, satisfaction, and anxiety in first language (Spanish) and second language (English) scientific writing for junior and senior faculty at a teaching university.   *77*

Figure 5.4 Means (figs. 5.4a & 5.4b) and standard deviations (fig. 5.4a) for self perception ratings on questions of difficulty, satisfaction, and anxiety in first language (Spanish) and second language (English) scientific writing for junior and senior faculty at a research institute (RI).   *78*

Figure 7.1 Progression of learning events through higher education and professional life (n=16).   *112*

Figure 7.2 Progression of learning events through higher education and professional life by subgroups (n=16).   *122*

Figure 7.3 A schematic representation of the valuable educational experiences that develop the ability to write research articles in English as a second language.   *125*

Figure 8.1 Three explicit teaching modules and related subsections.   *141*

Figure 8.2 Explicit teaching Content Delivery Matrix.   *142*

Figure 8.3 Types of knowledge that different types of associates bring to face-to-face interactions.   *147*

Figure 8.4 Suggested procedure for Faculty Writing Circle meetings.   *153*

Figure 8.5 Development of educational program.   *157*

List of Tables

Table 5.1 Summary of written statements made by participants concerning the difficulties they face in writing a scientific article in English (n=93).   *74*

Table 5.2 Percentage and frequency of writing strategy options for faculty at the teaching university and the research institute.   *75*

Table 5.3 Summary of degree of difference for self reported ratings on questions of difficulty, satisfaction, and anxiety in first language (Spanish) and second language (English) scientific writing for junior and senior faculty at a teaching university (TU) and a research institute (RI).   *79*

Table 7.1 Frequency of occurrence of scientists expressing anxiety or risk-taking concerning publishing scientific articles in English (n=16).   *130*

Acknowledgments

The research study that forms the centerpiece of this work began as a conversation between David and Karen on an autumn afternoon on the "tourist strip" of Ensenada, Mexico. In developing this conversation into a project, Karen worked closely and depended greatly on colleagues, administrators and students at the university and the research institute where she worked. We wish to acknowledge the Universidad Autónoma de Baja California for funds allowing David to travel to Ensenada on several occasions for the development of this study. We also thank David Toledo, Dagoberto Hernández, and Magdaleno Aviles for providing Karen with release time to work on this project. The quantitative survey instrument was translated by Carmen Márquez. The immense task of transcribing sixteen in-depth interviews was good-naturedly and efficiently performed by two wonderful colleagues: Carmen Yáñez and Claudia Vega. Assisting Karen with translation—particularly with idiomatic phrases that exceed her knowledge of Spanish—were two students who inevitably answered "help please" requests on instant messenger at all hours of the night; they are Xochitl Barney and Leticia Navarro. We thank Brian Fotinakes, Dr. Hayat Messekher, Eda Basak Azizoglu, and Cheryl Sheridan from Indiana University of Pennsylvania and Javad Bourbour Shirazi of York University for their help in proofreading and formatting this book.

This project could not have been completed without the trust and confidence granted to us by several influential scientists. David thanks Graham Hatfull for his continued support and advice throughout this project. We thank Eugenio Méndez and Roberto Millán for their personal work in supporting this research effort. In particular, we deeply thank Oscar Sosa who has gifted Karen with unscheduled meetings, strategy sessions, and key introductions that made this work possible. It is his personal commitment to the issues that we raise and his desire that we provide real solutions that inspires our work.

We extend thanks to the 153 anonymous scientists who were generous enough to participate in our survey. In addition, we wholly thank the sixteen scientists who shared their time, stories, successes, failures, and expertise with us when they have so much else to do. Their candidness always reinforced for us the importance of this work. We only hope that we can give back something that will ease the burden of writing professionally in a foreign language.

Finally, Karen wishes to acknowledge the friendship of Margaret Hogan and Kevin O'Donnell who listened attentively to her through excitement, frustrations, and fears during several years of this project's development. They are among a dear circle of friends in Ensenada, all of whom offered her their enthusiasm and support in whatever form it was needed, whether over a beer, a glass of wine, mariscos, or street tacos. Thank you.

Acknowledgment of Funding

David Hanauer was funded through a sub-award (IUP RI Log No.: 0910–028) from a Howard Hughes Medical Institute (HHMI) Professorship award to Graham Hatfull. This award also paid costs for the expedited editing of this book. Research work for this book was supported in part by internal research competition funds of the Universidad Autónoma de Baja California, Mexico awarded to Karen Englander. The authors thank both funders for their support.

Scientific Writing in a Second Language

1 The Dominance of the English Research Article in the Scientific World

1.1 English Literacy and the Generation of Scientific Knowledge

The world of science is characterized by the continued movement towards enhanced understanding of natural phenomenon, expressed through a range of communicative options that are predominantly based on literacy. In this sense, the world of science is dependent on literacy for the creation, dissemination, and preservation of knowledge. In recent decades, scientific publishing has become dominated by the use of the English language (Ammon, 1998, 2006; Hamel, 2007; Tsunoda, 1983); however, scientific research is conducted across the world by scientists for whom English may not be a first language. It is both a fair assumption and a claim that has some empirical support that second language scientists are at a disadvantage when it comes to publishing in English (Flowerdew, 1999b, 2007; Guardiano, Favilla, & Calaresu, 2007; Hanauer & Englander, 2011; La Madeleine, 2007). The difficulties faced by scientists for whom English is a second language when publishing their scientific knowledge in English poses a hindrance and a challenge to the continued generation of scientific knowledge.

The issue is primarily one of the diversity of scientific knowledge generation; the more diverse the background and contexts of scientists, the greater chance there is of generating new directions in thinking about phenomena of scientific interest (Hamel, 2006). Human diversity also translates into intellectual diversity. The enhancement of scientific knowledge benefits from the participation of as wide a group of scientific researchers as is possible, and it is unthinkable that quality scientific knowledge would go unknown because of difficulties in overcoming scientific publication. There is no doubt, from the perspective of scientific knowledge generation, that language and literacy should not be a barrier to the propagation of scientific knowledge. We are in agreement with John Flowerdew (2001), also quoted in Swales (2004:44), who states that "In a globalizing world, to place NNSs [Non-Native Speakers of English] at a disadvantage when it comes to publishing their work not only goes against natural justice but is also likely to be impoverishing in terms of the creation of knowledge" (p. 122).

From the perspective of the writers of this book, the issue of publishing scientific knowledge in English as a second language is an issue that should be of central concern to the world of science and not be seen as the problem of individual second language scientists or specific national, institutional, and linguistic groups. The issue of publication by second language scientists is an issue for science (and ultimately for humanity), as this addresses the conduit through which all scientific knowledge is currently communicated.[1] As stated by Robert Kaplan (2001), "Good scientists who cannot write English to meet the standards of journal editors are deprived of the opportunity to have their views and contributions disseminated," and by being excluded, "their contributions . . . are lost to science" (p. 18).

It is against this backdrop that the current book is written. Put simply, the aim of this book is to alleviate the barrier of English literacy when second language users of that language publish scientific knowledge. Our approach is multimethod and informed by both first and second language approaches to literacy instruction. Our interest in this issue resides in our respective academic working environments. One of the authors of this book (Karen Englander), works directly on English language, scientific writing issues, and research policies with Spanish speaking scientists in two research centers in Mexico; the other author (David Hanauer) works in an internationally staffed microbiology laboratory in the U.S., exploring the role of literacy and multimodal representation in the development and propagation of scientific knowledge, assessment, and education. Our work in these settings has convinced us of the significance of the scientific endeavor, the importance of English literacy within this context, and demonstrated for us the difficulties faced by scientists who are not first language English speakers.

To enhance the possibility of alleviating the barrier of English literacy from the publication of scientific knowledge by second language researchers, this book has three interrelated aims and modes of operation. First, the book aims to provide a comprehensive meta-synthesis of what we currently know about the phenomenon of second language scientific publication and the ways in which this issue has been addressed. The analysis appearing in different chapters in the first and last sections of this book covers areas of research into research articles written by first language writers, issues of second language writing—specifically L2 science writing—and offers a critical analysis of the pedagogical solutions posed so far. The second aim of this book is to present additional, qualitative and quantitative data on the phenomenon and problems faced by second language scientists publishing in English. Specifically, the research presented here aims to quantify the burden of second language science writing and explore the writing issues and historical solutions that have worked for specific second language scientists.

The third and most important aim of this book is to provide a framework of educational resources that facilitate informed, innovative approaches to alleviate the barrier of English literacy from publishing scientific knowledge by second language English writers. Our approach in posing a solution is based on three sources of information: meta-analysis presented within this book pertaining to the nature of the problems faced by scientists whose first language is not English, and the solutions proposed and documented within the literature concerning this population; our own research presented within this book concerning the nature and perception of the problems faced by second language scientists; and finally, our own experience working with literacy education in both first and second languages, and in relation to scientific writing. Broadly, as applied linguists who work with scientists, we believe that educational solutions should be based on the current state of knowledge, quality research, and creative and innovative thinking. In this book we aim to provide you, the reader, with an informed understanding and approach to address the needs of publishing second language scientists, and thus enhance the presence of scientific knowledge produced by researchers whose first language is not English.

1.2 English and the Scientific Research Article

In a recent book exploring the rhetorical power of science, Michael Zerbe (2007) characterizes scientific discourse as

> discourse that describes empirical research using the familiar IMRAD (Introduction, Methods, Results and Discussion) organizational scheme. For the most part this discourse is disseminated in peer reviewed, scientific research journals that are read chiefly by scientists who work in the specialty area on which the journal focuses. (p. 20)

He further points out that scientific discourse "has, since its inception in western Europe just three hundred years ago, become one of the most powerful discourses in Western society" (p.20). Within scholarly journals, the research article is widely positioned as the primary method for disseminating knowledge created in scientific disciplines), and is considered the principle means for individual scientists—and whole disciplines—to generate a corpus of shared, valued knowledge (Bazerman, 1988; Gross, 2006; Swales, 1990, 2004.

The importance of publishing within scientific journals is evident in sheer numbers. Today, there are over 100,000 scientific journals in the world, and 5,000 articles are published every day (Martel, 2001). Twenty years ago, approximately 66% of scientific articles were published in English (St. John, 1987), but now that figure has grown to 89% (Martel, 2001). Ammon (1998) places this figure slightly higher, stating that 90% of all scientific publications are in English. Figures on the growth of English scientific publication mark a very clear and rapid trend towards monolingualism in international, scientific publications (Hamel, 2007). Numerical data on both the number of scientific publications and the overriding dominance of English as the preferred language of scientific publication create a situation in which English literacy is plainly a prerequisite for participating in the international endeavor of generating scientific knowledge.

The impact of monolingual practices of English literacy in science publication goes beyond the question of whether a particular scientist will or will not be published. Whole communities of scientists who are publishing research in a first language other than English are having their research marginalized through linguistic discrimination. The problem is that the dissemination and evaluation of scientific knowledge is incomplete because work published in other languages is largely excluded from academic databases. The scientific community relies on academic, searchable databases to find valued scientific knowledge. Yet overall 74.57% of all articles, found within different databases and from a range of disciplines, are in English (Hamel, 2007). For specific databases, such as Pysch Info, the number of articles written in English can be as high as 95.2% (Cindoc, 1999).

Thomson Reuters Web of Science is, arguably, the most prestigious database that covers all scientific disciplines. It is used as the basis for most scientometric data, including rating the impact factor of journals and tracking the number of citations that accrue in articles. Language of publication is a factor that provides a distinct advantage for journals included in this database because Web of Science "focuses on journals that publish full text in English or at the very least, their bibliographic information in English" (Testa, 2012, para. 13). Thirty-six languages of publication are included in the database (King, 2004); however, its policy also states that "going forward . . . the journals most important . . . will publish full text in English." Since Web of Science provides a basis for evaluating the impact (in other words, value) of a journal or a specific article, and since one criteria of value is the presence of full text in English, journals and journal articles published in a language other than English are, by definition, marginalized and the dissemination of their scientific knowledge is limited and its value questioned.

Perhaps more importantly, the dominance of English in scientific journals and databases obscures the active publishing of scientific work in non-Anglophone countries. There are some 30,000 peer-reviewed journals published annually worldwide. Only one-quarter of them constitute the 9,500 journals included in the Web of Science (Salager-Meyer, 2008). Two percent, or 166 of the journals included in the Web of Science emanate from Latin America, Spain, or Portugal (Carlos Santiago, personal communication, November 9, 2007). Even so, many Latin-American journals bear English names and publish their work in English. In 2007, the Web of Science initiated a new program of regional representation to include 700 more journals from around the world. The language of publication among the majority of these regionally focused journals is English. Similarly, Ulrich's Periodical Directory indexes more than 52,000 scholarly journals; 74% of these are published in English (Lillis & Curry, 2006a). The domination of English holds true for specialized databases as well. GeoRef provides

"2.5 million references to the geoscience literature of the world," but includes less than one quarter of its references to articles *not* in English, and only 2% in Spanish. Similarly, the Database of Scientific Materials, which claims to provide access to "international research material rarely held in American libraries," includes only 2.6% of its references to Spanish serials. It is quite clear both in policy statements and in numerical evaluations that journals that publish in English, regardless of their country of origin, are more likely to be included in international databases than those that report on scientific knowledge in any other language.

This situation puts enormous pressure on scientists whose first language is not English to publish their research in English. On a basic level, the data presented above describes a system that privileges widely disseminated knowledge in high value publishing outlets. The underpinning prerequisite for this is full text publication in English, and to reach a broad and international audience, scientists publish in English to assure greater distribution of their work through academic databases. Their home institutions, fellow scientists, funders, and national science organizations evaluate the scientists' standing and status based on quality publication, which increasingly means publishing in scientific journals in English. This form of publication is considered the only significant way to contribute to the development of scientific knowledge within their respective disciplines.

This system of evaluating the quality scientific research has convinced academic institutions in many parts of the world to develop incentive programs that specifically promote publishing in English rather than in a national language. One form of incentive is merit pay awarded to individual scientists. The amount of the merit pay is tied to the number of articles published, the national or international scope of the publications, and the indexing of those journals in a database. Publishing in an indexed, international journal carries the highest value (CONACYT, 2009). Further, a country's national intellectual productivity is measured annually and reported by such bodies as the Organisation for Economic Cooperation and Development (OECD) and the World Bank. This intellectual productivity is measured in part by statistics generated by the Web of Science concerning the number of publications and citations generated by the nation's scientists. Nations have developed policies and programs to encourage participation in the international scientific arena, again emphasizing the importance of English based on English literacy as a prerequisite for the dissemination of professional scientific knowledge (La Madeleine, 2007; Uzuner-Smith & Englander, 2011).

The outcome of these processes is pressure to publish scientific findings first and foremost in English. This pressure originates from multiple sources: the individual scientist's desire to reach the widest possible audience and have the highest possible value assigned to her or his work; the institution's aspiration for international and national recognition as a site of research and its reciprocal offer of merit pay and other benefits to scientists who publish in internationally indexed journals; and national policies and programs that reward both institutions and individuals who produce recognized scientific publications. Nations are in fact under pressure themselves in that their standing in the international community and their access to assistance from international bodies is in part determined by their nation's intellectual productivity; this is defined in part by the number of articles indexed and citations accrued in databases that prioritize full text English journals. Ultimately, the situation is one that increasingly demands that an active scientist publishes her or his research in scientific journals in English. To change a well-known saying from the realm of academic survival, one could say that for the active scientist, no matter what your first language is, you either publish in English or you perish.

1.3 The International World of Science

One might assume that most scientific activity takes place only in English-speaking countries, and that the dominance of English-language publication is a reflection only of the quality of research within these English-speaking countries. However, this is not the case. For example, there are almost four million researchers working in the 30 countries that are within the Organisation for Economic Cooperation and Development (OECD). Of these 3.8 million, only 1.7 million work in the U.S. and U.K. (OECD, 2009). The 15 European Union countries of this organization have one million researchers. The final 1.5 million researchers are based in China, Japan, Korea, Poland, Mexico, and Turkey. Entirely excluded from the OECD figures are scientists in Africa, most of Latin America, and the Arab world. In another discipline-specific example, in the field of geophysics, 48% of the members of the Society of Exploration Geophysicists live outside the

English-speaking countries of the United States, Canada, the United Kingdom, Australia and New Zealand. Of the non-Anglo members, almost one-fifth reside in countries designated by the World Bank as disadvantaged (J.J. Paull, personal communication, April 25, 2003). Looking at a specific country, Mexico, for example, has 13,000 scientists distributed in all disciplines who are accepted into its annual national certification program for researchers (SNI, Sistema Nacional de Investigadores). Inclusion is limited to those who possess a doctorate and actively participate in their field through research, publication, and teaching (CONACYT, *Consejo Nacional de Ciencia y Tecnología*, 2009). Around the globe, 24% of the world's scientists are located in developing countries (Gibbs, 1995) where largely the national language is not English (India is an exception).

It is important to note that regional journals and non-English journals can make important contributions to the conversations in a discipline. They offer intellectual opportunities to discuss ideas within a nation such as Japan, Poland, or China, or across national boundaries that share a language, such as Spanish or Arabic. However, often these journals face various obstacles, and the scientific knowledge they generate is lost. Their readership is small; the quality of the science may adhere to a range of different criteria (Salager-Meyer, 2008); their distribution is often limited and publication schedule irregular (Canagarajah, 1996). The outcome is that these journals are not cited within other articles or wider databases, and as such, the knowledge they generate is unknown, ignored, and marginalized.

To bring the importance of local knowledge into perspective, it is worth considering cases where scientific research is directed at specific regional problems. For example, as stated by the editor of the British medical journal *Lancet* (as cited in Gibbs, 1995), at the time when concern about the Ebola virus was emerging in the West, research from local scientists in Africa on its microbiological characteristics and effects on populations was central in developing an understanding of this scientific phenomenon. In other cases, local knowledge or regional problems is not always of interest to scientists working within the Western world. Consequently, scientists who focus on issues of local and regional concern outside the Anglo world often encounter obstacles in publishing in American and British-based international journals. These journals may question the relevance of a local research question to their broad audience. Regional or non-English-language journals may publish the work, but such journals are often seen as peripheral, and the research they publish is condemned to having little visibility. It seems that scientists have little option but to publish in English and on issues construed as significant to an English-speaking world.

1.4 THE CRITICAL-PRAGMATIC APPROACH TO SECOND LANGUAGE SCIENCE WRITING

The writers of this book recognize that the situation of second language science writing is unjust and potentially harmful to the continued and diversified construction of scientific knowledge. A vicious cycle is created for scientists outside the Anglo world. To address critical issues of health, poverty, and the environment, it is crucial for scientists to be able to talk together. When they publish in their own language or in regional journals, they lessen the possibilities of reaching outside their immediate circle. Their contributions are limited by being less accessible. By being less accessible, they are not often cited. By not being cited, a scientist's work is not known, and the journal's impact factor suffers. As impact factor suffers, prospects for continued publication are weakened. This, too, lessens the likelihood of being indexed in the international databases that allow researchers to access scientific findings. Regional and non-English journals are not always good publishing choices, and a scientist's contributions therein fall into the domain of lost science.

The real question then is what can be done to alleviate this situation. It is clear to us that the situation of second language scientists writing for publication in English is one that has a critical component and to a certain extent needs a critical response. Previous research that we will review in greater detail in the coming chapters has established that scientists who are required to publish their scientific research in English as a second language face difficulties (Ammon, 1998, 2006; Flowerdew, 2007; Hanauer & Englander, 2011; Salager-Meyer, 2008). As we stated earlier, this situation is simply unjust and unhealthy for the development of science. However, this critical stance needs to be explicated somewhat. In a lecture on the teaching of second language writing for academic publication, Paul Matsuda (2008) pointed out that when a teacher works from a critical orientation designed to allow the second language writer greater access to their own second language voice that may be different from the standardized assumptions of editors and readers, there may be a price to be paid by the second language writer. John Flowerdew (2007) summarizes this in relation to scientific writing when he states that second language science writers need to be informed that "publication in local languages might be detrimental to career development, insistence on the validity of certain language re-use practices might give rise to accusations of plagiarism, insistence on writing in non-standard English might lead to rejection by editors and reviewers" (p. 23).

In responding to the needs of the second language science writer, Flowerdew (2007) proposes what he calls a critical-pragmatic approach. His position is that there should be a difference between the scholarly and practical work of a second language writing specialist. In relation to scholarly work, a clear, critical approach should be taken that contests "the inequities of the status quo" (Flowerdew, 2007, p. 23). However, on a practical level, the second language science writer needs to be helped to be able to publish their scientific research.

In this book, we adopt a critical-pragmatic approach. Our critical approach is contextual in that we believe that the situation is unjust, that the second language writer and his or her first language English writing peers, editors, reviewers, funding agencies, national and international science institutions need to be aware of this situation, that accommodations need to be made and that multilingual publishing options enhanced. But we also believe that the individual scientist and the field of science as a whole cannot wait for systemic changes in the current practice of linguistic discrimination. The answer we ultimately promote in this book is a framework of pragmatic support for second language science writers. Four kinds of educational resources are presented that separately or in combination can improve the writing success of scientists. We believe in locally determined, continued and institutionally situated extended support and co-writing with second language science writers. We believe that the need to publish quality scientific research in English is not the sole responsibility of the individual scientist and that the proposed solution should not rest solely in the intensive language education of that scientist. But rather this is a responsibility that should be shared by the scientists, the research center, the publishing outlets, funding agencies, and national and international science institutions. The solutions we propose in this book require resources, and part of the critical approach promoted in this book is directed at convincing various research centers, funding agencies, and national and international science institutions of the importance of this financial support. Ultimately, we wish to promote change by describing and promoting pragmatic educational resources and informed writing-support approaches that can provide second language scientists with the support they really need in order to publish research in English.

1.5 THE DESIGN OF THIS BOOK

This book has three sections. The first section (Chapters 1, 2, and 3) provides a theoretical introduction and a current literature review of research relevant to the discussion of second language science writing. The review addresses the genre specific aspects of the research article, studies of scientific writing by second language scientists, and socio-political aspects of English publication of scientific research. The second section of the book (Chapters 4 to 7) provides empirical data from the research project conducted for this book. The methodological aspects of the study are presented followed by the quantitative and qualitative results. The study provides a basis for the proposal of educational guidelines outlined in the third section. This last section of the book (Chapters 7 and 8) outlines a framework for educational approaches to the teaching of scientific writing to second language scientists. The educational resources are modeled on scientific processes of learning. Four elements are presented that can be modified and contextualized for local contexts: classroom teaching, modified writing center collaborations, expert and peer collaborations, and translation and editing services. This final section provides specific educational recommendations that facilitate enhanced levels of second language publishing in English.

2 The Genre of the Scientific Research Article

2.1 A Genre Approach

A basic principle of a genre-based understanding of literacy is the idea that specific types of texts evolve to fulfill and address communicative needs within a specific, social setting (Hanauer, 1998, 2004; Kamberelis, 1995; Miller, 1984). A text of a particular type fulfills social functions. Furthermore, the assumption is that genres evolve and change in structural and procedural ways in order to fulfill the dynamic needs of active communities of practitioners. For example, as documented by Gross, Harmon, and Reidy (2002) in a thorough study of scientific writing in three languages over four centuries, the scientific research article has changed and evolved dramatically from its beginnings as a form of personal testimony shared among gentlemen of an elite social status in European countries. The current chapter examines the scientific research article and aims to provide an informed, state-of-the art description of what is known within the language sciences concerning this genre. This discussion draws upon research conducted within the disciplines of applied linguistics, the sociology of science, composition, and the rhetoric of science, and is organized according to the tri-part description of a genre as a regularized literacy practice involving specific social functions, processes of production, and structural features.

2.2 Social Functions: Situating the Scientific Research Article

The prevalence and power of scientific discourse in the 20th and 21st centuries have brought us to the point where it is mundane or perhaps even trivial to state that the research article is a genre of social significance (Zerbe, 2007). It is not an exaggeration to state that the research article is the major measure of scientific success and is crucial to the livelihood, credibility, and status of any scientist (Hyland & Salager-Meyer, 2008). Basically, a scientist cannot be a scientist without producing and publishing research articles. Bruno Latour and Steve Woolgar (1986), in their well-known sociological description of laboratory functioning, state that the "production of papers is acknowledged by participants as the main objective of their activity" (p. 71). Thus, on a simplistic level, it could be argued that the publication of the research article is the aim of scientific activity.

But what does the scientific research article do? In other words, what are the social functions of this genre within communities of scientists? The most obvious answer to the first question is that the research article disseminates scientific knowledge. In this sense, the research article is a tool for communicating and sharing scientific developments between different interested parties. Rhetoricians and sociologists of science have established the idea that this "dissemination" is not the impartial presentation of objective knowledge, but rather a rhetorical act of persuasion within the context of competing conceptualizations of the world (Bazerman, 1988; Latour & Woolgar, 1986; Harris, 1997; Myers, 1990; Prelli, 1989). Randy Harris (1997), in his overview of rhetorical case studies of science writing, situates this understanding of the rhetorical nature of science in the seminal work of Thomas Kuhn's (1962) analysis of paradigm shifts in the sciences. Harris (1997) argues that a scientific claim is only meaningful within the social context of the specific orientation and approach of particular communities of scientists. This focus on the community positions rhetoric and persuasion at the center of the scientific enterprise and describes the nature of scientific activity as the process of generating support for particular types of claim.

Latour and Woolgar (1986) situate scientific work as the production of a series of inscriptions and documents designed to allow the social construction of facts. In making their case, they differentiate five levels or types of statement: Type 5 claims are "taken-for-granted" statements that are so well-known to experts in the field that they are rarely discussed but rather assumed to be known as basic facts; Type 4 claims are similar to Type 5 in that they are well-established, but these claims are explicitly stated, such as in introductory scientific textbooks as factual statements; Type 3 claims, often found in a review of previous literature, add modality to the stated fact, thus questioning, to a certain extent, the factual nature of the claim; Type 2 claims are far more tentative and offer thoughts on what might be reasonably concluded from the evidence on hand; and Type 1 claims are speculative and consist of conjecture. These five types can be seen as a continuum that moves from "fact-like entities" (Type 5 claims) to purely speculative assertions (Type 1 claims). In this formulation, the role of a laboratory is to allow a process in which Type 1 statements presented in the lab can be moved up the hierarchy of claims. As argued by Latour and Woolgar (1986):

> The name of the game was to create as many statements as possible of *type 4* . . . the objective was to persuade colleagues that they should drop all modalities used in relation to a particular assertion as an established matter of fact, preferably by citing the paper in which it appeared. (emphasis in original, p. 80)

The aim of a laboratory is to change the status of statements from Type 1 to Type 5, and the aim of the research article and all of its structural features is to allow statements to be made that persuade others that they are of the standing they need to be considered factual, and hence, will be subsequently referenced in other papers.

On a functional level, the process of referencing another research article involves a professional interaction with previously published research. The explication of this interaction reveals the rather complicated web of literacy within which science takes place and the primary role that the research article plays in this construct. Eugene Garfield (1965) specifies the following reasons why research articles are cited:

1. Paying homage to pioneers
2. Giving credit for related work (homage to peers)
3. Identifying methodology, equipment, etc.
4. Providing background reading
5. Correcting one's own work
6. Correcting the work of others
7. Criticizing previous work
8. Substantiating claims
9. Alerting to forthcoming work

10. Providing leads to poorly disseminated, poorly indexed, or uncited work
11. Authenticating data and classes of fact-physical constants, etc.
12. Identifying original publications in which an idea or concept was discussed
13. Identifying original publications or other work describing an eponymic concept or term.
14. Disclaiming work or ideas of others (negative claims)
15. Disputing priority claims of others (negative homage). (p. 189)

While some of these usages deal with recognizing and assigning value to another scientist, many are related to the reasoning behind the way a study was done, why it was done, and the content of the claims being addressed. In this sense, the research article provides information that becomes part of the conceptual process that constructs and directs new scientific research. The production of research articles and their subsequent citation creates an intertextual network of related texts and claims that constitute previous scientific knowledge. As such, research articles play a significant social role in allowing access to established and existing thoughts and understandings on a particular phenomenon. Furthermore, this knowledge base provides the context within which scientific argumentation and the proposal of specific claims take place. Scientific arguments take place in relation to existing scientific claims, which are referenced through the presence of published research articles.

However, it is a mistake to think that the research article is purely functional. As with other social systems, publishing a research article with its associated scientific claims is tied to a socially constructed hierarchy of value. As seen in the writing cases analyzed by Myers (1990), higher level claims were initially directed at more general and widely read science journals with enhanced social value; but, following rejection, more specific claims were directed at disciplinary and sub-disciplinary journals with a more focused and limited professional readership and a lower social value. The choice of publishing outlet dictates the exposure and readership of a published article and the possibility to which this article will be later referenced and cited in subsequent publications. The aim of a researcher is to have their scientific claims be widely referenced and cited, which would serve the double purpose of disseminating information and increasing self value as scientists.

This issue of the social hierarchy of publications has quite direct symbolic, professional, and economic benefits, which is why scientists place such high importance on publication (Latour & Woolgar, 1986; Franck, 1998, 1999). In this sociological analysis, knowledge generation and dissemination is subordinate to the process of generating self value and opportunity through publication. Latour and Woolgar (1998) describe this as a cycle of credit in which publishing that does not directly produce monetary benefit can be translated, through a credit process, into enhanced value as a scientist and a professional. The increased status of the scientist following publication facilitates additional opportunities for funding and professional development, thus creating a cycle in which publishing enables advancement in other areas of scientific activity.

Georg Franck (1998) sees the process of citation—explicit reference to the published work of a scientist—as a process that directs attention towards a particular researcher; thus enhancing that researcher's reputation. Publication and subsequent citation, in this analysis, is analogous to the manufacturing output of a scientist and is a direct measure of their productivity value. As in Latour and Woolgar's (1998) analysis, for Franck (1998), publication and citation translates into institutional promotion, enhanced chances for funding, and additional publication options. Thus, publishing a research article fulfills what can be called a socio-economic function; it assigns symbolic social value to the publishing scientist and provides that scientist with economic benefits in the form of employment and research funding.

The social value of publication and citation is evident in the late 20th Century development of a series of bibliometric measures of journal and article publication. These measures offer quantitative data on the degree to which a given article, researcher, institution, country, or continent produces research articles that are published and subsequently cited. Measures such as *productivity* (basic counts of papers produced), *efficiency* (a ratio of the number of citations for a particular paper, or group of papers, that offers an analysis of the impact of the published research), and *aggregate citation counts* (compares the number of citations for researchers over a period of time, and thus offers a comparative measure of the impact and influence of published research) offer quantitative ways in which the value of research is evaluated (Pendlebury, 2008). It is important to note the central role the research article plays in this process, as all of these measures are tied to the publication of articles and to the internal citation within these articles of other research papers. These quantitative measures of research publication are increasingly used in evaluating a scientist's value.

The social function of a research article is not, however, limited to the individual researcher or laboratory. Institutions, cities, countries, and continents are evaluated for their scientific value through citation counts and publication productivity. For example, in a recent report in *Science,* it was found that 19 universities in the U.S. produced 47% of all the citations from the U.S., and beginning in 2008 the total output from the U.S. fell below the output of the combined 27 countries of the European Union and below a conglomerate of Asian-Pacific countries (Mervis, 2010). Thus institutions, countries, and continents are compared using bibliometric data based on article publication. Bodies such as the Organisation for Economic Cooperation and Development (OECD) and the World Bank measure the intellectual productivity of countries and regions based on metrics of articles published, and they make educational-financial decisions based, in part, on this data. Thus, the published research article has become the major component of the evaluation of research value of a person, institution, or geographical region. So while the most obvious functions of the research article might be the dissemination of knowledge, the establishment of existing claims, and the direction and development of new scientific research, an analysis of the social role of the research article reveals that it is a measure of individual and group scientific and intellectual value that facilitates economic and professional opportunities.

2.3 Producing a Research Article

The process of producing a research article is tied to two very different contexts—research within the laboratory (or in other research settings) and interaction with publishing outlets. Writing a research article is an extended process of several months (to over a year) that integrates what was done as research and the requirements and responses from journal editors and reviewers. Broadly, the contours of this process consist of interaction in the research setting over the construction of a research article and then interaction with specific comments of reviewers and editors over the status of a paper and how it can be revised. As described by Myers (1990), a large part of this process is attaining the appropriate level of claim for the study conducted. This movement over the generalizability or specificity of the claim is part of an extended negotiation between the authors and the type of journal and its specific journal editors.

Myers (1990), in his description of the process of writing a research article by two different researchers, specifies the following stages: (1) a draft of the paper was written by each author and not submitted for publication; (2) a more focused paper was written and directed at a major journal; (3) following rejection of the paper by the major journal, each paper was rewritten and resubmitted with a cover letter to the same journal; (4) following a subsequent rejection by the same journal, the papers were revised and submitted to another major journal with a wider readership; (5) following rejection, the papers were rewritten for a more specific audience and sent to journal with a smaller, more focused readership; (6) following rejection, the papers were revised and resubmitted to the same journal and were ultimately accepted for publication. This long and tedious process makes clear the degree of social interaction involved in the publication of a research article, including multiple reviewers and editors interacting with the authors of these research articles. This is an inherently social process of writing.

Both Myers (1990) and Knorr-Cetina (1981) in an earlier analysis of the research article writing process note that a major part of the revision process consists of changing the modality of certain statements and making arguments stronger by deleting sections and reshuffling others. In the detailed analysis of changes offered by Myers (1990), it is clear that the scientists were careful to follow the requests of the reviewers and editors, but were reluctant to do so. The specific changes imposed upon them consisted of making the papers fit the more conventional structure of the research article and changes to a range of self-positioning statements. Arguments were tightened, and data were added to the papers so that the clarity of empirical arguments could emerge. Of course, as has already been stated several times, the modality of the claim was adjusted so that it conformed with what could be said based on the data presented. The process of writing a research article can be characterized as an extended social process involving extensive and quite sensitive understanding of the status and aims of each journal as well as the specific requests of editors and reviewers.

There is an additional aspect to the process of writing a research article that has been elegantly captured in a recent book by Theresa Lillis and Mary Jane Curry (2010). The extent of the social contextualization of the research article writing process goes beyond just the interaction between researchers, peer-reviewers, and journal editors directed by the abilities of an individual scholar. As described by Lillis and Curry (2010), a network of resources is actually required for the research article to move forward. These resources can include both "language brokers" who address the linguistic aspects of an English-medium publication and "academic brokers" who help negotiate knowledge construction, production, and dissemination. Fundamentally, the actual process of producing a research article in English is a networked process involving multiple actors. As such, the process of writing a research article is complex from a social perspective, and success depends on a range of required interactions with different agents within this network of resources. It would be simplistic to think of the writing of the research article as situated only with the abilities of the individual scientist.

2.4 The Structural Features of a Scientific Research Article

In conjunction with the increased solidification of English as the language of science, over the 20th century, the structural features of the research article have become largely standardized (Gross et al., 2009). As discussed above in relation to the social functions of the academic article, the social and economic value of this literacy product have meant that there are enormous pressures in making individual writers, reviewers, journal editors, and publishers conform to a standardized set of valued textual and stylistic characteristics in their manuscript preparation. The outcome, as clearly shown in Gross et al. (2009), is a process in which certain characteristics of the research article have become increasingly consistent. This relates to the organization of the research article, its linguistic features, and its argumentative structure. The overall direction of the standardization of the research article can be summarized as a movement towards the succinct expression of informational complexity, depersonalization through the foregrounding of knowledge statements, and facilitating navigational clarity (Gross et al., 2009). These sets of features are discussed individually below, but it is already important to note that the development of these textual characteristics suggest a development in which finding specific information quickly by experts in a particular field is the main aim of this way of designing a manuscript. The scientific research article is user-friendly to the expert-insider who comes to the reading with specialized knowledge and has specific informational aims in reading a particular article. However, while describing these features, it is important not to forget, as is reviewed in the next chapter, that these features and the process of standardizing the research article are not just functional, but rather fulfill ideological roles that define groups of users and the values assigned to them. Scientists not following this standardized set of linguistic and textual features face the risk of having their science marginalized (Burrough-Boenisch, 2003; Li & Flowerdew, 2007).

2.4.1 The Succinct Expression of Informational Complexity

As documented by Gross et al. (2009), one core development in the research article from the 17th to the 20th century is that:

> grammar has adapted by adding substantially to the complexity in its noun phrase and by the deployment of specialized literary devices (such as fused noun strings and abbreviations) aimed at compactly conveying technical messages to small groups of highly trained readers in a specialized research field. (p. 167)

Scientific writing in English in the 20th century is characterized by the increased usage of complex noun phrases with multiple modifiers in both subject and non-subject positions. The complex noun phrase in scientific discourse is constructed by placing multiple modifiers to the left of the central noun. This linguistic construct allows the expression of complex, information-rich utterances in a compact form. While this also increases difficulty and potential ambiguity for novice readers, for the expert-insider, it allows quick access to already acquired concepts. In this sense, complex noun phrases function as a form of abbreviation. For the informed reader, this way of presenting information allows complex thought to be expressed and handled with far greater ease, because the complex noun phrase becomes a single, conceptual entity that is easily comprehended. The exact opposite is true for the novice reader, who would need to literally decipher the various elements of the noun phrase and relate them to real world correlates, greatly increasing the complexity of reading.

A parallel development in English scientific writing in the 20th century is the increased usage of both explicit abbreviation and citation conventions. As with the complex noun form, both of these aspects of scientific writing allow the expert insider direct and swift access to required information. As with the process of complex noun phrases, for the insider, well-known abbreviations and citations provide a depth of intertextual relationships to be expressed succinctly and processed with relative ease. For the novice, both abbreviations and citation conventions involve a lengthy process of decoding and reconstruction. Complex noun phrases, abbreviations, and citations all function as forms of a signaling and referencing process for informed insiders, allowing swift access to relevant information.

The increased usage of complex noun phrases with multiple modifiers, abbreviations, and citations is augmented by a different development in English scientific writing—the decreased complexity of sentence structure. As documented by Gross et al. (2009), the overall length of the English scientific sentence is 28 words and 2.2 clauses per sentence. These numbers are down from previous centuries, in which sentence length was measured as 33 words in 1876–1900, 30 words in 1901–1925, and 27 in 1976–2000 (Gross et al., 2009). Clause density remained relatively constant throughout, at approximately two clauses per sentence. Thus, the linguistic data on scientific writing presents an interesting picture of increased noun phrase complexity and usage of abbreviations and citations within a decreasing sentence length and consistent clausal density. These features of English scientific writing create a construct especially designed to present complex information succinctly to experts. It is a structure that functions mainly through processes of quick access and referencing, and allows expert readers to express and comprehend complex information in a compact linguistic form.

2.4.2 Depersonalization through the Foregrounding of Knowledge Statements

A long recognized and moralized aspect of writing scientific research articles is the presence of passive voice. This aspect of scientific writing has been attacked for its construction of a sense of impersonality (Stratton, 1984; Weinberg, 1967) and for removing and obscuring agents of knowledge construction (Hansen, 1998; Williams, 1999). Broadly, from a critical perspective, the assumed problem with the linguistic construct of passive voice in scientific writing is that it creates the illusion of fact and objectivity while obscuring the person constructing knowledge (Hildebrand, 1998). Moreover, the use of the passive voice has been criticized for obscuring the characteristics of scientific inquiry as it hides both researcher and the obstacles faced by that researcher in conducting their inquiry (Hoffman, 1988). The use of passive voice in scientific writing has also been attacked for distancing the audience and for not making scientific writing more accessible to the public (Mair & Hundt, 1995).

On a more functional level, the presence of passive voice in scientific writing has been explained in terms of a shift from a focus on the person creating knowledge to an interest in knowledge itself. For example, scientific writing with the passive voice has been described as "instrumental and object orientated" as opposed to "actor-orientated" (Salager-Meyer, Defives, & Hamelinsck, 1996). Alan Gross (1990) has stated that scientific writing avoids "any device that shifts the reader's attention from the world that language creates to language itself as a resource for creating worlds" (p. 43). In other words, as argued by Gross (1990), scientific writing aims to foreground claims constructed in language and makes these claims the center of the reader's attention. The argument for passive voice in scientific writing is that the focus is on evaluating the knowledge claims and the evidence presented in support of those claims. As seen in the historical data presented by Gross et al. (2009), the presence of passive voice is part of a larger construct that foregrounds "technical language and painstaking description" and reduces personal language, such as evaluative expressions and poetic metaphors and similes (p. 166). This type of writing situates knowledge claims and their appraisal at the center of the scientific reading and writing processes.

However, the outcome of the criticism of passive voice and other depersonalizing aspects of scientific writing has been the claim that passive writing is decreasing (Seoane, 2006). Anja Wanner (2009), in her research, argues with this claim by entering into a more nuanced understanding of active and passive voice. Building upon Jaeggli's (1986) argument, Wanner (2009) has stated that passives include an implicit agent, and as such, cannot be seen as agent-less. Furthermore, active voice does not actually mean the explicit presence of an agent. Structures such as "These numbers show that . . ." or "This paper argues that . . ." function in the same way as a passive construct from the perspective of foregrounding the knowledge claim and backgrounding the researcher as an agent. Wanner's (2009) analysis presents two patterns through which knowledge is foregrounded while using active voice. She terms these: "Fact Construction" (e.g., "The structural data demonstrates . . ." or "These results contradict . . ." (p. 179) and "Paper Construction" (e.g., "This article suggests . . ." or "This paper argues that . . ." (p.179). Thus while her data does show a decrease in the use of passive, it also shows an increase in the use of fact and paper as agents, which still obscures the researcher. These results are interesting in that they reveal an underlying aim in scientific writing to focus on knowledge and knowledge evaluation. This aspect of science writing complements the structural features of complex noun phrases, citations, and abbreviations that allow experts quick access to relevant information. The whole construct of fore grounding knowledge claims and making highly specific and complex knowledge accessible in the most compact forms would seem to be designed to allow experts the ability to properly and quickly access information and evaluate its value.

2.4.3 Facilitating Navigational Clarity

In discussing the development of the ways in which conceptual complexity in scientific writing is handled at the phrase and sentence levels, it was pointed out above that increased usage of complex noun phrases, abbreviations, and citations create a linguistic construct that mainly functions as a compact referencing system. This movement towards ease of access to information for expert readers is also present in the standardization of the textual form of the research article. Two interrelated developments in the textual features of the research article facilitate this movement towards ease of access and clarity of navigation: the requirement for a specific progression of categories of information, and the requirement that each category be clearly marked with appropriate headings and subheadings. Together, these two features create a textual structure with conventionalized expectations for specific types of knowledge and an easy way to find required knowledge at marked textual sites.

The overall organization of the experimental research article has been well-documented, and is widely recognized as consisting of the *Introduction-Methods-Results-Discussion* (IMRD) format solidified through Swales' (1990) analyses. Each of these macro-components of the research article has a specified communicative role and discursive construction. The *Introduction* section of the research article has received the most attention from a rhetorical perspective. Swales (1990), in an influential analysis, proposed understanding the introduction in relation to the *Create a Research Space* (CARS) model. This model specifies three basic moves that position the described research: (1) *Establishing a territory*, which involves reaffirming the importance of the field of inquiry; (2) *Establishing a niche*, which involves specifying inadequacies within existing research, whether this involves a lack of research or an issue with the existing research; and (3) *Occupying the niche*, which involves providing a justification for conducting the research presented in the rest of the paper. This structure can be seen in terms of a problem-solution schema in which, initially, a field of interest is charted, and then a problem is defined to which the article and research is a solution. Within science, the type of problem that is usually solved is a well-defined, smaller problem within a very specific conceptual and empirical frame. The construction of the solution in the third stage of the CARS model explains the purpose of the research article.

The *Methods* section of the research article fulfills an important role within 20th and 21st century science, replacing what in previous centuries was an evaluation of the writer's authority based on status. The methods section establishes the credibility and authority of the study itself by explicating the way in which the research presented in the article was conducted. Thus, the methods section has less to do with the replicability of the study itself, and more to do with the establishing the credibility of the specific research plan used to address the problem described in the introduction. Accordingly, while the methods section does provide extensive detail on the techniques, procedures, and materials used in the study, the detail is not enough to actually replicate the study exactly without further consultation with the researchers (Gross et al., 2009). As opposed to previous centuries, in which the description of method was a form of personal narrative outlining the exact steps taken, current methods sections tend to be disembodied without mention of actors, and includes formulaic and conventionalized references to previously documented research procedures, materials, and tools (Gross et al., 2009; Swales, 1990). As seen in other sections, the method section can be easily read by the expert-insider who would be familiar with the particular methodological references being made, and could easily comprehend and infer what was actually done by the researcher.

The *Results* section of the research article is designed to provide readers with the outcomes of the presented methods. Not all the raw data can be presented, so researchers must reduce the data to an appropriate size for presentation. Here, figures and tables are often used as summary devices. A second step in presenting results is generalizing from the data (Penrose & Katz, 2010). As noted by Swales (1990), this section is written in a way that suggests that the results are agent-less (or at the very least, the agent is not a human being). Swales (1990) presented several examples of this style, such as "The ANOVA on . . . indicated," or "Tukey post hoc tests indicated that . . ." (p. 170). This linguistic strategy is described by Wanner (2009) as fact construction in which the result itself is used as the subject of the sentence. However, there is some evidence that different disciplines may present results with different components. As reported in Swales and Feak (2004), in a study of results sections in biochemistry, several different types of commentary could be present within a results section, including justifications of methodology, interpretation of results, agreement with previous studies, evaluative comments concerning the data, admitting difficulties, and pointing out discrepancies. This suggests the potential for a much wider discussion within the results section in relation to the results themselves. However, as pointed out by Swales (2002) in relation to Williams (1999), medical research involved very little in the way of commentary.

The *Discussion* and *Conclusion* sections of a research article have a cyclical relation with the *Introduction* section. The findings are discussed and interpreted in terms of how they extend, refine, or challenge previous findings or assumptions (Olsen & Huckin, 1991). Gross et al. (2009) argue that the conclusion section consists of three components:

> (1) original claims derived from having occupied the niche or solved the problem defined in the introduction, (2) wider significance of those claims to the research territory, and, (3) suggestions on future work to validate or extend the original claims. (p. 179)

This three-part structure re-emphasizes the researcher's claims and situates them within the current research. Hopkins and Dudley-Evans

(1988) provide a wider set of rhetorical moves that may be present in the discussion. These include referencing background information, making a succinct statement of results, addressing unexpected outcomes, citing previous research, providing explanation, exemplification, recommendations, and through deduction and hypothesis, generalizing the original claims. Thus, a range of strategies can be used to conclude the research article, the most important of which is the provision of a clear statement of the researcher's knowledge claims and their significance.

The IMRD structure is situated within a marked textual frame that once again fulfills the function of foregrounding specific knowledge and allows easy navigation. These elements include the title, name and affiliation, abstract, and a list of citations and acknowledgements. The presence of citations and acknowledgements situates the research within a net of intertextual and professional relations, and addresses the community within which particular research is valued. The title and abstract in the 20th century research article are designed to provide quick access to the main themes and claims of the research article. As analyzed by Gross et al. (2009), 30% of titles present the researcher's main claim, and 65% present the main theme of the paper without explicitly stating the main claim. The abstract provides a succinct summary of the whole study, including purpose, method, results, and interpretation of the results. The author's name and affiliation offer some insight into the standing, position, and physical location of the researcher. Together, the title, abstract, and author information offer a very quick introduction to the knowledge claims and research conducted by the researcher, and as such, offer the reader a way of deciding whether the article is worth further attention. Thus, the initial framing of the research article further enhances the ease of navigation of the whole article.

Overall, the structural features of the research article create a text that is rich in information, has conventionalized expectations for positioning and finding particular types of information, and functions mainly through a process of referencing. This type of writing is quite obviously designed for a specialized group of readers with expert knowledge in a particular field, and foregrounds the development and evaluation of knowledge. In addressing the linguistic features of the research article, Gross et al. (2009) state that "these measures have helped improve communicative efficiency, in partial compensation for the growing conceptual and semantic complexities of the subject matter and purposeful narrowing of intended audience" (p. 172). The direction of the research article over its historical trajectory has produced a text type that allows experts to quickly and easily navigate and find information they seek. For the expert-insider, it is a technology that provides a compact, conventionalized delivery system of highly specific information in an ordered structure; for the novice-outsider, it is almost impossible to follow because of the conventionalized form of complex noun phrases, abbreviations, and citations.

2.4.4 Persuasion and Stance in the Research Article

As discussed in relation to the social function of the research article, persuasion and acceptance of the associated status of a knowledge claim (and by extension, the researcher) is a central aim of the research article. Genre theory posits that structural features facilitate social function, and accordingly, in this section, the way the research article persuades its readers and situates its claims and authors is discussed. Crucial to this process is the idea that scientists, in writing their research articles, are attempting to position themselves and their claims as supported by the evidence presented. On the one hand, scientists need to accurately position the statements they make in relation to the evidence they present through a process of careful linguistic contextualization and qualification, termed hedging. As stated by Hyland (2000a), this hedging process consists of modifying "the epistemic warrant of their claims" (p. 93) through the assignment of a series of hedges to the statement itself. These hedges can be of various types, such as epistemic reporting verbs—"suggest" and "indicate"—or adverbial and adjectival lexicon, such as "clearly," "possibly," and "minor." In this way, a statement can increase or decrease its authority, prescriptiveness, explicitness, assurance, assuredness, and power. The type of hedge assigned to any statement determines the status of the knowledge claim ultimately made. The specific hedge used needs to reflect the type of inference that will be accepted by a community of scientists in relation to the data and research presented.

On the other hand, as described by Hyland and Salager-Meyer (2008), there are interactional aspects of the research article. These involve constructing the appropriate authorial voice and relations (Hyland, 2005a, 2005b). Broadly, for an argument to be persuasive, the scientist-author needs to present a written persona that will be recognized and accepted by other researchers. This means recognizing the disciplinary context and the ways others have addressed the issues at the center of the research paper. Hyland and Salager-Meyer (2008) differentiate between what they call *stance*, "the writer's textual voice or community recognized personality; the ways writers present themselves and convey their judgments, opinions and commitments" (p. 307), and *engagement*, "the way writers acknowledge and connect with their readers, pulling them along with their argument, focusing their attention, acknowledging their uncertainties, including them as discourse participants, and guiding them to interpretations" (p. 307). Stance and engagement construct a textual space within which an argument concerning the status of a knowledge claim can be evaluated.

As seen in the review of the structural features of the research article, this text type is designed to deliver experts easy access to a series of knowledge claims. Of course, the main social aim of this writing is to convince scientists within a particular field of the importance of presented research. In other words, the aim is persuasive and involves convincing readers to accept the writer's argument. The scientific argument in the 20th century is characterized by the centrality of method, the reliance on measurement, quantification for the establishment of fact, the comparison of data sets, the presentation of mechanistic explanation, and the use of visual representations to succinctly present large amounts of data in useable formats (Gross et al., 2009). The argument of a scientific research article is based on the presentation of a methodological design that is deemed appropriate by the expert reader to provide evidence that can answer the question posed. In science, this method often includes comparing data sets using appropriate methods of measurement and quantification presented in table or graphic form. The stated outcome of this process involves establishing fact and a mechanistic explanation. If the method is appropriate, and has been operationalized correctly, if the measurements and findings are accurately, succinctly, and clearly presented, and if the claim is appropriately qualified, the overall argument should be accepted by an expert reader.

2.5 The Research Article

As seen in this brief review of the genre features of the research article, this text type is a highly specialized and developed technology for the transmission of particular information to expert readers. The structural features of the research article, with its clear textual level, navigational features, its foregrounding of knowledge statements, and succinct expression of information, make it ideal for allowing experts to quickly and easily reference and evaluate the quality of the information presented in the article. Broadly, every element of the research article—from its title, abstract, explicit marking of organizational sections, its use of graphs and tables, complex noun phrases with multiple modifiers, abbreviations, and citations—are all designed to provide the expert reader with two core functions: easy navigation and quick referencing of complex information. This type of writing is highly utilitarian and designed for a particular audience, whose aims are to acquire and evaluate the presented knowledge. At the same time, and on a broader social level, the research article clearly plays a role in evaluating the intellectual value of an individual scientist or group of scientists. Enhanced value through publication brings with it both economic and professional opportunities. Thus, while the aim and structure of a research article may be the dissemination and evaluation of knowledge, the personal aspects of social value, professional advancement, and financial success are directly related to the publication of research articles.

3 Second Language Writing and the Research Article

3.1 Introduction

A scientific research article presents knowledge that has been vetted by referee-peers and presented in an accepted written form. An article is constructed to disseminate this knowledge and enlist peers in the acceptance of that knowledge. It is "a rhetorically sophisticated artifact that displays a careful balance of factual information and social interaction" (Hyland & Salager-Meyer, 2008, p. 305).

The constructed nature of scientific research articles has been thoroughly discussed in the previous chapter. The fact of its construction can present challenges for two types of scientists: scientists who are geographically located outside the resource-rich countries of the U.S., Canada, and the U.K., and scientists who do not speak English as their first language. The obstacles are even greater, then, for a non-native, English-speaking scientist who is based in a periphery country. The purpose of this chapter is to synthesize the research on writing science in a second language and publishing science from the periphery.

The research informing this chapter is drawn from a number of fields, including applied linguistics, second language writing, English for academic and/or special purposes, and the sociology of science. The data is organized into several different categories addressing the socio-economic and geographic context, peripheral participation, interacting with the journal editors, difficulties and strategies of writing, and contrastive approaches and language community characteristics. Overall, the aim of the chapter is to extend the last chapter's discussion of the research article by contextualizing this genre within the world of second language scientists who are distanced from the resources required for science.

3.2 Socio-Economic and Geographic Context

The world of research is not conducted on an equal playing field. Scientific research is conducted in a wide range of contexts that have disparate resources and privileges. In trying to understand and categorize this disparity, a series of terminological distinctions have been developed. Dominant, resource-rich countries termed "center" have been distinguished from resource-poor countries which are designated as "peripheral" sites of research (Wallerstein, 1991). Similarly, based on economic and industrial development, sites of scientific research have been defined as First World versus Third World, developed or industrialized countries versus developing countries, and rich versus poor (Salager-Meyer, 2008).

Furthermore, the English language, the most predominant medium for expressing scientific knowledge, is also not an equally shared resource. Braj Kachru (1996, 1997), in a well known analysis of the status of English, distinguished three "circles" of countries based on their relationship to English. "Inner Circle" countries are those where English is the dominant first language in all domains, and is comprised of the U.S., U.K., Canada, Australia, and New Zealand. The "Outer Circle" is comprised of former colonial countries such as India, Nigeria, and Singapore, where English has an official status for matters of government, but is not the first language of the large majority of the population. "Expanding Circle" countries are those where English is a foreign language but may be used in restricted domains, such as education or commerce. This includes most of Europe, Latin America, Asia, and African nations that are not former British colonies. It should be noted that European nations have much greater economic advantages and long traditions of scientific endeavors in contrast to Latin America, for example.

By themselves and in isolation, the terminological concepts proposed so far fail to capture the variation of privilege and position that exists in conducting and reporting science (Lillis & Curry, 2010). Accordingly, we differentiate between three terms that address the linguistic and economic sources of power in relation to scientific publication and research. The term "Anglophone countries," which includes the U.S, U.K., Canada, and Australia, designates research sites that have both economic and linguistic privileges. The term "center countries," which include Anglophone countries as well as Western Europe and Japan, designates countries that have economic resources for science, but are not necessarily L1 English-speaking countries. The term "peripheral countries," which includes Latin America, Africa, and most of Asia and the Middle East, have both linguistic and economic disadvantages. This terminological scheme involves evaluating both the linguistic and economic resources dedicated to the scientific activity of a given country, and basing the definition on the state of these two variables.

The economic and linguistic advantages of Anglophone countries are exemplified through the fact that one third of the scientific articles published in a given year come from scientists in the U.S., and 30% of the world's research and development (R&D) dollars are spent in the U.S. (Kuznetsov & Dahlman, 2008). In contrast, Mexico, a peripheral country, which is one of the two poorest nations that are members of the Organisation for Economic Cooperation and Development (OECD), spends about a quarter of one percent of the world's dollars on R&D, and produces three-quarters of one percent of published articles.

Furthermore, the economic advantages of center countries is seen through bibliometric data on scientific publishing used for most national and international comparisons and published by Thomson Reuters Web of Science (formerly known variously as Thomson ISI, the Institute for Scientific Information and the Science Citation Index). King's (2004) study on the scientific impact of nations reports that 32 countries, including the G8 and E.U.-15, account for 98% of all highly cited papers, and the remaining 162 countries (all peripheral) account for less than 2%. King's (2004) data is based on the Web of Science, which collects and reports data on 8,060 journals (available at: http://wokinfo.com/products_tools/multidisciplinary/webofscience/), including 700 recently added "regional" journals.

It should be noted that the economic and linguistic advantages of Anglophone and center countries are further enhanced through the way scientific success is measured through entities such as Thomson Reuters Web of Science bibliometric data. For example, in this corpus, the journal master list represents only one quarter of the world's peer-reviewed journals. Frequently, national journals are excluded from the database even though national journals "play an important role in the dissemination of research in locally-oriented areas, such as Earth Science or Clinical Medicine" (Bordons, Fernández, & Gómez, 2002, p. 203). As a result of excluding many national journals, "the quantity of production of these [peripheral] countries . . . is frequently underestimated" (Bordons, et al., 2002, p. 199).

Journals that are produced in peripheral countries face a range of obstacles that limit their inclusion in the Web of Science, from erratic publishing schedules to a lack of international editorial board members, to publishing in a language other than English (Salager-Meyer, 2008). It is important to note that journals are not required to publish in English to be included in the Web of Science, although they must provide an abstract in English. Nonetheless, the overwhelming majority publish entirely in English (95% of the natural science journals and 90% of the social science journals, according to Lillis and Curry, 2010, p. 9). The emphasis on English language publication is repeated by the Web of Science Director of Editorial Development, who writes, "Going forward, it is clear that the journals most important to the international research community will publish full text in English" (Testa, 2012, para. 13). Also, as pointed out by Lillis and Curry (2010) and others, the term "international" in the context of academic publishing is often used as a proxy for "English medium." Peer-reviewed journals that do not publish in English, and are produced outside of the developed countries, are largely excluded. Thus, the scientific work they contain is simply not counted in national or international biometric data. Consequently, scientific knowledge produced in non-Anglophone countries may simply not be counted.

The economic challenges of peripheral countries go beyond the absence of dedicated funds for research and development, and address the more basic aspects of being an academic. Lack of material resources as basic as sufficient paper, postage, and reliable postal services were noted 15 years ago (Canagarajah, 1996, 2002a). Today, reliable power sources for electricity, computer, and Internet access are fundamental for academic publishing. Without these material resources, access to the author guidelines for submission to individual journals may be limited, and knowledge of bibliographic and documentation conventions may be unfamiliar. To position research in the larger conversations of the discipline, access to the current literature through full-text data bases, journals, and monographs is crucial. Where these non-discursive resources are lacking, scholars have difficulty preparing their manuscripts. These obstacles are distinct from the linguistic presentation of a manuscript, and yet they serve to limit the possibilities of a manuscript being accepted for publication.

Finally, dominance of English in the realm of scientific publishing accounts for two additional trends of the geoeconomics of publishing: the marginalization of publishing in national languages, and a nation's greater ability to participate in English language publishing. For example, bibliometric data from Spain indicates that in the ten-year period of 1996 to 2006, there was a 52% decrease in articles published in Spanish and a corresponding 51% increase in articles written by Spanish researchers in English (based on figures in Pérez-Llantada, Plo & Ferguson, 2011). This "demise of Spanish-language journals" (Pérez-Llantada et al., 2011, p. 22) offers fewer outlets to scholars who only speak Spanish. The ability to function in English, as calculated on a national scale, was found to be the determining factor in a 2004 study of published medical research articles (Man, Weinkauf, Tsang, & Sin, 2004). A country's TOEFL (Test of English as a Foreign Language) score is more directly correlated to scientific article production for the 20 member countries of the OECD than even the dollars invested in research and development. Access to high levels of English proficiency within a non-English-speaking country has a significant effect on international publishing success.

The economic status of a nation can have consequences for the research and development possibilities available for scientific endeavors. Disadvantages accrue to scientists in peripheral contexts because of lack of material resources. Furthermore, a nation's English language proficiency is a determining factor in international journal publishing. Finally, when scientific knowledge production is considered, most bibliometric figures exclude the substantial publishing that occurs in national and regional non-English journals. In other words, what gets counted as knowledge production is limited. These issues minimize the value of scientific work and publishing that takes place outside of Anglophone and center countries.

3.3 Peripheral Participation

Conducting research from the periphery involves experiences that are qualitatively different from those of scientists situated at the center. The literature has addressed this through a number of approaches: the so-called "exile from paradise"; research activity networks as a key entry into formal and informal collaborations; and the imbalance of collaboration between peripheral and center scholars. Each of these approaches highlights different aspects of the experience of researching and publishing that involves scholars outside the Anglophone and center countries.

The phenomenon of physical relocation by international scholars who study in the Anglophone center and then return to their own country has been described through the metaphor of "exiled from paradise." This phrase is used by Swales (1990) and Flowerdew (2000), but comes from Geertz (1973), who says "one starts [an academic career] at the center of things and then moves toward the edges" (as cited in Uzuner, 2008, p. 258; also discussed in Salager-Meyer, 2008). This metaphor, which might address the economic privileges of center countries, also has obvious colonial overtones with its assumption that paradise is the Anglophone center, and that any leaving must have been exile. Undoubtedly, some international scientists may wish to locate themselves in North America or Britain, but others are equally anxious to gain their education and return home. A few recent studies have explored peripheral scientists' purposeful determination to be located in their own countries and publish in their own languages (Salager-Meyer, 2008; Flowerdew & Li, 2009), especially to contribute to their local knowledge base in order to address problems of local concern (Duszak & Lewkowicz, 2008; Curry & Lillis, 2010). It is interesting, however, that the notion of being exiled has been adopted to capture the idea that marginalization awaits the scholar who leaves a center location.

When a scholar is located in the periphery, connections with the center are valuable in achieving success in English-language, international, publishing. Lillis and Curry's (2010) recent book, *Academic Writing in a Global Context*, is based on an eight-year study with 50 scholars in the social science fields of psychology and education in 12 institutions in four European countries. By tracing "text histories," their findings demonstrate the central importance of networks in understanding non-Anglophone scholars' success or difficulty in publishing in English-language, international journals. Lillis and Curry's research (2010, 2006a, 2006b; Curry and Lillis, 2010), which emphasizes networks, challenges the notion of individual competence as key to publishing success. They argue that the focus placed on English language skills is insufficient to explain the "real-life text production practices of scholars" (2006b, p. 63). Networks are what actually explain the peripheral scholar's publishing success.

Networks can be categorized as local or international ("transnational" in their nomenclature), and as weak or strong. Different networks provide different resources, but crucially, international networks "can provide information, opportunities, and support for English-medium text production, leading to publication" (Lillis & Curry, 2010, p.86). This success is accomplished largely through contact with academics who work in Anglophone-center contexts, termed "academic brokers." These academic brokers provide peripheral scholars with knowledge that shapes their texts through conversations concerning the specific content, discipline, and target journal possibilities. Through such interactions, peripheral scholars can learn of publishing opportunities and shape their texts in ways that are more likely to be accepted by journal editors. These academic brokers help to shape the content of a text, such as indicating what to emphasize, how to position knowledge claims, and how to interact with journal editors. Note that these interactions with academic brokers are distinct from the kind of interactions that may take place with "language brokers." Language brokers attend to sentence level or linguistic content of a text and are usually different people from academic brokers. Nonetheless, networks that include brokers are "highly desirable, if not essential. This is particularly important for gaining access to Anglophone-center-based information and the crucial linguistic and rhetorical resources needed for English-medium high status publishing" (Lillis & Curry, 2010, p. 69). While peripheral scholars are not characterized as exiled in Lillis and Curry's view— since they may never have studied in the Anglophone center—they have much more favorable potential to establish and enhance networks if they began their careers with graduate or post-graduate work in the Anglophone center.

One way for peripheral scholars to network with center scholars is through research collaboration. International collaborations in the hard sciences and engineering, however, can be especially complex because of the social context of collaboration. Three prominent factors determine those contexts: funding, scientific actors' networks, and interdisciplinarity (Hwang, 2008; Katz & Martin, 1997). These factors can be understood as internal or external to science (Wagner & Leydesdorff, 2003). Internal to science factors are disciplinary differentiation that leads to interdisciplinary and multidisciplinary efforts, as well as the tremendous amount of equipment and resources required to conduct "big science" (Galison & Helvy, 1992). Factors that are external to science and that create the context for international collaboration include public and policy support for research and development, historical relationships, colonial ties, geographic proximity, and access to communication technologies such as the Internet.

From the point of view of the peripheral country, several advantages accrue through collaboration with a center country's institutions. These advantages include access to state of the art research equipment and prestige for the peripheral scientist when he or she returns to the home country. However, peripheral scientists point out that this access seems to be bought:

> I think the only way of collaborating with advanced countries is by us [Koreans] offering more funding to international collaborative projects. We pay expenses for inviting foreign researchers, and we pay when we visit foreign laboratories. The primary purpose is to get advanced knowledge and technology transferred. . . . From the point of view of advanced countries, when they collaborate for new knowledge production, they do not need us, because they can do it for themselves. (reported in Hwang, 2008, p. 126)

A consequence of the dichotomy of central and peripheral nations is the kind of scientific knowledge produced. Peripheral scientists, due to lack of infrastructure and their shorter scientific heritage, often consume scientific knowledge that is produced by scientists from the center. Scientists in the periphery produce subsidiary knowledge, and this dichotomy creates a division of labor in science production (Hwang, 2008, 2005). This imbalance of knowledge production in central and peripheral countries occurs in the social sciences as well: "Theoretical and methodological perspectives provided in the core theories of [a] discipline [are] delineated mostly by theorists and researchers in the Anglo-European centers (e.g. U.S., U.K.)" (Lin, 2005, as cited in Oda, 2007, p. 125). Academics, whether they are within or beyond the English-language centers, apply these "global theories" to local issues and places (emphasis in original, Oda, 2007, p. 125). For Korean scientists and engineers, "Collaboration between Korean research organizations and advanced organizations is not primarily under-taken for the purpose of knowledge production but, rather, that it benefits activities such as knowledge transfer, career building, model application to local conditions, or fund-raising" (Hwang, 2008, p. 126).

So, what accrues to the core country through such collaborations? The benefits are working with those who may be the most talented scientists in a peripheral country, having data processed in a context that does not use an advanced country's resources, and exploring results applied to new contexts. A further consequence of this unequal division of labor between center and peripheral countries is that claims of new knowledge production from a peripheral country "may be prejudged in light of [an] assumed lower status" (Hwang, 2008, p. 106). This assumption that profound knowledge does not get produced in the periphery can affect how journal editors and reviewers respond to manuscripts from peripheral scholars that make such claims. (See the next section for further discussion.)

Connecting with members of the center of the scientific community, when physically located outside the center, is a powerful advantage. Canagarajah refers to this as developing "the social practices that surround the scholarly text" (personal communication, May, 2010). Such connections with core members was identified as important in Diane Belcher's (2007) investigation, of which papers written by peripheral scholars get published in one applied linguistics journal. Research network theory (Lillis & Curry, 2010) demonstrates the crucial importance of Anglophone center connections, even for those non-Anglophone scholars in Europe. These connections may be formal collaborations or informal acquaintances; they may begin with graduate study in an Anglophone country; and they are not the sole means of achieving international publishing. The research does indicate that peripheral location is concurrent with greater difficulty in publishing a paper; however, in itself does not necessarily doom these scholars to exclusion.

3.4 Interaction with the Journal Editors

Decisions about which manuscripts are accepted for publication are made by journal editors with input from reviewers. The reviewer's goal "is to ensure that inaccurate or sloppy research is weeded out" (Yaffe, 2009, p. 1) but also to assure that "new discoveries . . . get disseminated to the scientific community as rapidly as possible" (Yaffe, 2009, p. 1). Once a manuscript is published, it is "up to the community at large to come to their own conclusions about the soundness of the research" (McGinty, 1999, p. 138).

Reviewers are generally selected by editors because of their expertise in the field (Gordon, 1980; McGinty, 1999). However, reviewers are not always neutral judges. Reviewers come from circles or networks that are known to the editor, beginning with their personal networks of institutional affiliations and then those of their colleagues. In the five most important medical journals, for example, only two of the 111 editorial board members come from "low income countries" (Salager-Meyer, 2008, p. 127). The overwhelming dominance of Anglophone researchers as journal editors creates a situation where manuscripts submitted by researchers in the periphery are reviewed less favorably. It is important to note that in the natural sciences, authors and their institutional affiliations of a manuscript are identified to the reviewers, and this practice is distinct from social science fields, which practice double-blind review. This identification disadvantages scientists situated outside those English-speaking countries because their institutions may not be known to the reviewers. In this situation, reviewers seem to bring a degree of skepticism to work that would not otherwise be applied and such bias or misgivings, based largely on an author's provenance, can precipitate a manuscript's rejection (Ceci & Peters, as cited in Kourilová, 1996; Link, 1998; Shashok, 1992; Wenneras & Wold, 1997). More than a dozen studies "provide clear evidence of bias favoring authors from the U.S., English-speaking countries outside the United States, and prestigious academic institutions" (Salager-Meyer, 2008, p. 125). In contrast, Sedef Uzuner's (2008) survey of the literature concerning the publishing activities of multilingual scholars cites two studies that refute the existence of bias (Li, 2002; Liu, 2004). Currently, empirical evidence is mixed in this regard (Salager-Meyer, 2008).

It is interesting to note that a number of editorials in diverse scientific fields have recently commented on the importance of supporting non-native English-speaking scholars and including submissions from beyond the Anglophone countries. For example, when Hugh Gosden (1992) asked editors if there is bias regarding such submissions, one responded, "Yes, bias in favour!" (Gosden, 1992, p. 130). In fact, some editors lament what they see as reluctance and lack of confidence of non-native English-speaking researchers to submit papers. The journal editors encourage submissions from all over the world to reflect a "broad readership and an international status" (Gosden, 1992, p. 130).

Studies that focus on the review process from the author's perspective tend to indicate a perceived discrimination in the evaluations they received based on their location outside the Anglophone center or their non-native English-speaking status (Englander, 2009; Curry & Lillis, 2004; Hwang, 2005; Kerans, 2001; Aydinli & Mathews, 2000; Li, 2002; Flowerdew, 2000, 1999a; Sionis, 1995). For example, a prominent Mexican marine scientist explains a particularly harsh criticism he received, "When suddenly someone from Mexico, unknown, wanted to throw in something that is important, [the reviewers] say, 'No, that has to be said by me'" (Englander, 2009, pp. 43–44). A Hungarian psychologist, who has published extensively in both his first language and in English, reports that "if the style or the form of the paper is not native . . . reviewers think that 'this is a stupid man'" (Curry & Lillis, 2004, p. 678). Interestingly, a Korean physicist with more than 200 published papers noted that journals editors were more critical of his manuscripts when he moved from the U.S. to Korea, stating, "Some of them sometimes made kind of silly comments on my English" (Cho, 2009, p. 236). He attributes this new scrutiny to his non-native status physically located *outside* of the Anglophone center.

Not all scholars perceive such bias. A recent study was conducted with Korean doctoral students who often were first-time published authors, and they downplayed the importance of English in reviewers' judgments about their manuscripts (Huang, 2010). The students believed that "good physics is good physics" (Huang, 2010, p. 38), and further, they expected that criticism of their writing was inevitable. These novice scientists believed that the content of a manuscript supersedes presentation in editorial judgments, and this belief is echoed by many established Korean scientists (Hwang, 2005). Similarly, some non-native English-speaking scientists report that they see the reviewing process as unbiased and they state that reviewers do not consider either the scientist's country or language in their judgments of a manuscript (Lillis & Curry, 2010; Matsumoto, 1995).

The reviews that journal editors and reviewers write to non-native English-speaking scholars have been the subject of a few studies. Three studies looked at editorial behavior specifically within fields related to applied linguistics: Flowerdew (2001), Belcher (2007), and Tardy and Matsuda (2009). John Flowerdew (2001) reports that the editors in his study stated that they were "generally satisfied with their reviewers' attitudes when dealing with [non-native English-speaking] manuscripts, in spite of the fact that it was not unusual to find insensitive comments in some reviews" (p. 145). Many of the editors "described cases in which they in fact went out of their way to help [non-native English-speaking] contributors" (Flowerdew, 2001, p. 129). Overall, the editors found that surface errors in language to be a lesser problem than issues of parochialism. Parochialism points to the non-discursive resources discussed above in which the authors do not demonstrate a good understanding of the current issues and developments in the discipline, and this can be due to lack of access to current bibliographic material in under-resourced countries.

Belcher (2007) examined 29 reviews of nine manuscripts (including resubmissions of revised manuscripts) directed at authors from outside English-speaking countries. Nine text features were most commonly remarked upon: topic, audience, purpose, literature review, methods or research design, presentation and analysis of results, discussion or significance, pedagogical implications, and language use or style. Her frequency counts indicated that 93% of the reviews commented on language, and while 6 reviews made positive comments, 90% of the reviewers made negative comments about language. Comments regarding language use and style were by far the most consistent category across all reviews.

More recently, Christine Tardy and Paul Matsuda (2009) surveyed 70 editorial board members of journals in language- and writing-related fields with a focus on reviewers' construction of author identity in the discipline's typical double-blind review process (i.e. the identity of the author is not revealed to the reviewer, and vice versa). One of those respondents remarked that novice writers often "self-reveal as being written by a non-native speaker based on an accumulation of language errors. I have rarely if ever recommended publication of these pieces that are written by novices" (p. 45). Non-native writers are not necessarily novice writers, but this reviewer conflates the two. Swales (2004) also stated that the difficulties of non-native English speakers are like that of novice writers. In his view, by moving from novice to expert writer, the difficulties of a non-native speaker similarly disappear.

A few studies focus specifically on reviewers' comments in fields other than applied linguistics. Gosden (2003) was also interested in what reviewers commented on regarding manuscripts by Japanese researchers in "a hard science field." As in Belcher's (2007) study, comments referred to technical details, claims, discussion, references, and format. In this instance, just less than half (18 of 40) of the reviewers commented on some aspect of language, and the majority referred to general problems such as "the authors should try to improve the English" (p. 99). This proportion is similar to the comments received by Italian medical researchers where 44% of comments were language based, and 56% were critical of scientific content or method (Mungra & Webber, 2010). However, isolating language-based criticism from scientific comments was often found to be difficult.

Fortanet's (2008) study of Spanish researchers in applied linguistics and business organization extended Kourilova's (1996) examination of the discourse structures of evaluative language in reviews. By classifying the structures as criticism, recommendation, and question, she shows how directives and other face-threatening acts are sometimes mollified. Their indirectness may in fact be difficult to interpret on the part of non-native English-speaking authors.

Most recently, Englander and López-Bonilla (2011) examined two sets of reviewer comments sent to two Mexican scientists. The reviewers' comments revealed their beliefs about what constituted "good science" and a "good paper" and, most importantly, whether the reviewer felt that the Mexican researcher shared that same understanding. When the reviewer felt the beliefs were shared, his or her attitude toward language difficulties in the manuscript was helpful; when the reviewer determined that the beliefs were not respected by the scientist-author, language difficulties were emphasized. Although the reviewers commented on the same manuscript, and all noted non-native English, some severely criticized the language, and others offered linguistic assistance. The authors posited that reviewers adopted one of three stances toward the scientist-author, with the consequence of advancing the Mexican scientist's manuscript toward publication or dismissing it.

When non-native English writers receive criticism, it can have a negative effect, according to a prominent language sciences journal. The editorial board of *TESOL Quarterly*, the premier American journal in its field, implemented a policy about ten years ago for reviewers of manuscripts written by non-native English-speaking authors. Paul Matsuda, a non-native English-speaking scholar and member of the editorial board recounts, "I had heard stories about NNEST [non-native English-speaking teachers] authors receiving reviewers' condescending comments about their language, and I thought it would be a good idea to ask reviewers to refrain from such comments" (P. Matsuda, personal communication, 2004). Consequently, the following sentence has been added to the letter sent out to manuscript reviewers: "Although it is often appropriate to discuss the way the paper is written, we prefer not to have reviewers comment negatively on the writer's control of English" (S. Canagarajah, personal communication, 2004).

To date, studies indicate that rarely are manuscripts excluded solely based on perceived shortcomings of language skills (Guadiano, Favilla, & Calaresu, 2007). Most alarming, however, is journal editors or reviewers who conflate poor language skills with poorly conducted science. A former editor of the prestigious journal *Science* is quoted as saying, "If you see people making multiple mistakes in spelling, syntax and semantics, you have to wonder whether when they did their science they were not also making similar errors of inattention" (Gibbs, 1995, p. 96). Such a statement confirms the perception of some scientists, as reported in the previous section, that their work is not evaluated seriously because the writing in their manuscripts reveals their non-native English-speaking status.

Editors do use reviewers' comments to reject a manuscript. The language can "be as good a reason as any" (Gosden, 1992) for rejection, especially when manuscript acceptance rates are low. The title of a 2003 article in *The Scientist* displays this sentiment with "No pardon for poor English in science" (Jaffe, 2003). The editor of *Biochemistry* is quoted as saying, "If a paper comes to me in poorly written English, I send it back. I don't want to waste anyone's time" (Jaffe, 2003, p. 45). A more sympathetic view comes from the editor of a physics journal that receives many international submissions. That editor reports that he would work directly on a manuscript "if I understand what the author is trying to say scientifically but I do not think my copy editors do" (McGinty, 1999, p. 37). The world of science journal publishing combines many pressures: perpetually insufficient time, a greater volume of submissions to each journal than is possible to publish, and editorial schedules that make demands on academics serving in volunteer capacities. These pressures all coalesce to "depict a struggle for recognition on the part of each submittor [sic]," (McGinty, 1999, p. 41) so that when manuscripts are submitted by non-native English-speaking writers, and they require more editorial resources to shape them into acceptable form, the chances of the authors receiving those resources are limited. Salager-Meyer (2008) summarizes the situation as follows:

"The problem is, firstly, that non-discursive factors very frequently go hand in hand with poor linguistic skills (at least in non-English-speaking periphery countries) and, secondly, that poor linguistic skills go hand in hand with paper rejection" (p. 125).

In summary, we see that editors and reviewers of scholarly journals have the power in deciding which manuscripts are published, with the attendant prestige and knowledge dissemination that publishing entails. There is some evidence that manuscripts produced by scholars in the non-Anglophone center and the periphery are subjected to a level of scrutiny or suspicion not applied to Anglophone scientists or those based in center institutions. Not all studies support this perception of bias. Most studies with journal editors and reviewers indicate that they identify problems with many manuscripts that non-native English speakers submit, and these are attributed to language and readability. In some cases, these manuscripts are quickly rejected. In other cases, there are editors who actively welcome peripheral scholars, and they work with those authors to improve the manuscripts when they perceive language shortcomings. Generally speaking, there seems to be relatively low tolerance for manuscripts that are perceived to be "non-native," and recommendations to authors to have their manuscripts read by a native English speaker prior to submission is commonplace.

3.5 Difficulties and Strategies of Writing

Non-native English speakers who want to participate in international conversations of scientific investigation have reported that they have difficulties writing their manuscripts in the foreign language, and in some cases, they develop strategies to overcome them. Studies have investigated the processes and techniques of writing science texts in English as a second language. In some cases, those processes and techniques are highly personal and idiosyncratic. In other cases, they have been reported as strategies shared by many writers. Those difficulties and strategies are examined here.

It is important to point out that there is substantial literature that addresses the process of enculturation and identity formation as practitioners of a discipline for university undergraduate and graduate students, particularly in the U.S. and the U.K. (Belcher, 1994; Fox, 1994; Ivanic, 1998; Li, 2005; Poe, Lerner, & Craig, 2010; Prior 1998; Tardy, 2005; Winsor, 1996). Students' sense of affiliation with or alienation from their field of study is one of the factors that impacts the degree to which they feel motivated to adopt the conventions of scholarly writing of their discipline. Mentoring from senior faculty is reported as an important factor in their sense of entitlement to participate in scholarly writing (Cho, 2004; Dong, 1996). Graduate education in an Anglophone center university can add to a scholar's confidence in writing in English (Flowerdew, 1999a), and scientists sometimes lament the fact that they did not do their graduate or postdoctoral work in an English environment (La Madeleine, 2007). Those studies that explore the socio-affective side of writing by students are put aside for the moment because interest is focused on the scientist who has completed formal education. Therefore, we examine the research concerning the writing difficulties and textual strategies reported by non-native English-speaking scholars.

The difficulties that scholars report are discussed in Uzuner's (2008) review of 39 empirical studies. Problems that can be described as not mastering the language code are common. These are problems that are "technical . . . with respect to having less facility of expression in English and vocabulary . . . convoluted syntax and unclear modality . . . and inappropriate or incorrect use of idiomatic expressions" (p. 255). A subsequent paper by Cho (2009) corroborates these findings for Korean scientists, where "linguistic elements" were the most troublesome.

Another commonly reported difficulty is the fact that writing is time-consuming, protracted, and tedious. Writing in a foreign language is slower in simply getting words on the page (Silva 1993; Flowerdew, 1999a; Curry & Lillis, 2004). The writing process is also prolonged because of repeated revisions to specifically address language concerns. Manuscripts written by non-native speakers of English are subjected to rounds of revision both prior to and post submission (Burrough-Boenisch, 2003), and still publication may not be successful (Li & Flowerdew, 2007).

Scholars also report rhetorical difficulties with writing different sections of research articles. Methods sections are usually considered easier to write as they are more technical in content, whereas introduction and discussion sections are more difficult (St. John, 1987; Swales & Feak, 2004). These sections require rhetorical skills to respond to the literature of the field, document sufficient evidence, and make claims of appropriate strength. These rhetorical sensibilities can be difficult to manage for a native speaker, and especially so for a non-native speaker of English (c.f. Uzuner, 2008). In part, the rhetorical difficulties are attributed to language, such that inadequate management of the language of confrontation, politeness, evaluation, boosting, and hedging results from insufficient knowledge of English. However, as will be discussed in the next section, these may be cultural differences that affect the rhetorical choices of non-native English-speaking authors. Thus, difficulties reported by second language scientists in their writing are both linguistic and rhetorical and the protracted nature of repeated revisions with a specific focus on language serve to interfere with their ability to share their scholarly work with international audiences in a timely manner.

The task of writing manuscripts for English publication is approached differently by different individuals. For some, extensive use of the first language preceding writing in English is common. For example, Spanish scientists (St. John, 1987) and Japanese doctoral students (Gosden, 1996) reported that they produce a complete text in their first language and then translate it into English. This translation is often performed "phrase by phrase" or in groups of two or three sentences (Gosden, 1996, p. 116). This strategy often preserves first language phraseology and organizational structure when the manuscript is reproduced in English. One Mexican scientist reported that when a manuscript was returned by a journal in part because of language problems, he rewrote the entire manuscript in Spanish and then hired a translator to put it into English (Englander, 2009). Scientists in St John's study found that having their papers translated from Spanish into English "proved unsatisfactory" (St. John, 1987, p. 116). Alternatively, scientists write an outline of their manuscript in their first language. Through this strategy, they are able to organize the flow, and it enables them "to explore ideas fully on their own intellectual and cognitive levels" (Kobayashi & Rinnert, as cited in Gosden, 1996, p. 115). The outline is used as the structure for their English manuscript.

Another strategy is to write notes or an outline in English (Gosden, 1996; Sionis, 1995), and the developing manuscript is sometimes supplemented by borrowing from other papers (Flowerdew & Li, 2007; St. John, 1987). From this skeleton, the complete paper is created. Scholars may borrow or lift words, phrases, and occasionally paragraphs from other people's papers in what has been termed "jigsaw" writing (St. John, 1987). When faced with transforming a text from a first language into English for publication purposes, Sionis (1995) found that it was not uncommon for scientists to simplify their English text relative to their first language French text. Their strategies included avoiding a topic or reducing the message present in the first language when their text was written in English. Spanish scholars in Spain also report they sometimes simplify their ideas because they lack linguistic skills (Pérez-Llantada et al., 2011). Certainly the comprehensiveness and subtlety of writers' ideas suffer when they feel they are unable to fully express their knowledge.

At the other end of the continuum of strategies, some scientists report that they write directly in English and do not rely on their first language (Matsumoto, 1995). In their self-perception, their first language does not play a role in their publication efforts. For example, one-third of the 585 academics in Hong Kong in Flowerdew's (1999a) study reported that they had no disadvantage relative to native English speakers in their publication writing practices, and two of the three highly prolific Korean scientists in Cho's (2009) study said they had no feeling of being disadvantaged. Studies conducted in the European Union with Danish, Swiss, and German scholars also report that they are not hindered by writing in English (Ammon, 2001; Murray & Dingwall, 2001; Preisler, 2005). In countries where English has been used as the medium for much university education, some scientists even report that writing scientifically in their mother tongue is troublesome (S. Camacho, personal communication, May 21, 2008) or may become so (Ferguson, 2007).

Language assistance is available to non-native English-speaking authors, both formally and informally. Scholars often report asking a native English-speaking friend or colleague to review their manuscripts, and this may or may not be greeted warmly. For example, one American chemistry professor no longer does so, saying, "It's too time-consuming to work on someone else's English writing skills" (reported in Jaffe, 2003, p. 44). In a collaborative team, the native English speaker is asked to pay special attention to the language of their manuscript (Englander, 2006). Paying for editorial services can also be an option. There is a growing industry of proofreading and editing through companies such as American Journal Experts (prices begin at $250 USD for editing a "short paper" of 500–1500 words, at http://www.journalexperts.com), Bioscience Writers ($70 per hour, at http://www.biosciencewriters.com/pages/prices.htm), and Academic Word. In addition, freelancers abound with varying abilities in terms of disciplinary knowledge, bilingual capabilities, and language expertise. Such "language brokers" (Lillis & Curry, 2006a) who aim to improve the texts of non-native English speakers are often sought out (Burrough-Boenisch, 2003). Non-native scholars' interactions with these brokers are still under-studied; however, of 300 scholars based at a university in Spain, 53% reported using them in their publication efforts (Pérez-Llantada et al., 2011).

In summary, the relationship between the first language and English in writing can be seen on a continuum. At one extreme, is the scientist who writes entirely in the mother tongue and hires a translator to put the text into English. In this situation, the scientist has almost no written production in English although refinement of the translated text may occur. At the other extreme, the scientist performs all the writing of the paper with its attendant cognitive and intellectual work in English. They report no conscious use of their first language. Between these two extremes are strategies that include outlining, phrasal borrowing, asking native English-speaking friends or fellow scientists to review, and hiring "language correctors." It must be noted that apart from those who feel no disadvantage in composing and writing in English, the demand to submit manuscripts in a second language slows down their writing and is often reported as being more difficult. This extra burden interferes with their scientific publishing productivity.

3.6 Contrastive Approaches and Language Community Characteristics

Recent decades have seen a substantial body of work about the language and cultural differences in academic writing. For example, Abdollahzadeh (2011), Pérez-Llantada (2010, 2007), Duszak (1997), Yakhontova (2006), and Ventola (1991) are among those who document how scientific texts, stances, and styles of intellectual debate are connected with cultural norms, values, and beliefs. Different theoretical lenses have been used to explore the different conventions of scientific writing in different languages, including genre analysis (Swales, 1990; Bhatia 2004), the sociology of scientific knowledge (Berkenkotter & Huckin, 1995), systemic functional linguistics (Englander, 2006; Martínez, 2001, 2003; Martín-Martín & Burgess,

2004), and corpus linguistics (Hyland, 2000b, 2008). In sum, these researchers have identified that there are variations in convention and that these differ in relation to the language of publication (e.g., see Englander (2010) for a detailed analysis comparison of Spanish and English textual differences). These variations are generally not well accepted when reproduced in English-language scientific manuscripts intended for publication. Hyland and Salager-Meyer (2008) present a comprehensive overview of research, and two key concepts are discussed here: the CARS model and reader- versus writer-responsible texts.

Swales's (1990) seminal work on the generic structure of the scientific research article described the series of rhetorical moves that writers perform in organizing their articles. His "create a research space" (CARS) model, discussed at length in the previous chapter, has generated much subsequent analysis and fine-tuning from the original presentation of establishing a research territory, establishing a niche, and occupying the niche. Research has shown that this is a specialized Anglophone style and it is difficult or even contrary to the style expected in other national or language contexts. Swales commented that Anglophone countries with their attendant "big world, in big field, in big languages, with big journals and big libraries" research articles need to reflect "academic promotionalism and boosterism" (Swales, 2002, p. 71). These elements may not be as welcome in academic communities that are smaller or peripheral. For example, Gosden (1996) reports Japanese reluctance to claim the research space through the establishment of a gap and avoidance of the confrontational ways of framing science articles (Gosden, 1996, p 122–123). Sudanese (El-Malik & Nesi, 2008) and Korean (Cho, 2004) scholars want to avoid perceived "boosting" or emphatic statements and the confrontation that Western rhetoric expects. In contrast, Mexican scientists can be hyperbolic or "exaggerated" (Englander, 2009, p. 43) and Hong Kong Chinese scholars can be "too assertive" (Flowerdew, 1999b, p. 256) in their claims when writing in English. Chinese scholars also tend to refrain from citing relevant texts that they consider "deficient or incomplete" (Hyland & Salager-Meyer, 2008, p. 317) because of cultural expectations of appropriate interaction. In the case of Spanish social scientists (Burgess, 2002) and Malaysian scientists (Ahmad, 1997), the small size of the community constrains scholars from making claims that might challenge colleagues whom they know and rely on for other collaborative and funding activities (Burgess, 2002). A statement that indicates the gap in previous research (i.e., Swales's move two) was found to be present in only 15% of Spanish-language abstracts (Martín, 2003), and this is explained as a consequence of the small community who would read the work in Spanish. "The reduced number of members belonging to this community makes it unnecessary for the Spanish researcher to establish a niche" compared to the members of the international academic community where "there is more competitiveness to publish and consequently a greater need to justify their work" (Martín, 2003, p. 41).

An important distinction in the creation of scientific texts is whether language is "reader-responsible" or "writer-responsible." The differentiation comes from Hinds (1987), who characterized the explicitness of cohesion and coherence in texts as being typical of different language styles. English is considered a writer-responsible language in that it is the writer's responsibility to clearly and coherently lay out the text for easy comprehension on the part of the reader. This is accomplished through a high degree of metadiscourse with explicit transitions, signals of coherence, and markers of logical unity. Other language traditions expect the reader to make sense of the text, where there are fewer explicit transitions or discourse markers. Writers in Asian languages such as Korean, Japanese, and Chinese expect their readers to "think for themselves and draw their own conclusions" (Hyland & Salager-Meyer, 2008, p. 320). Scholars writing in French, German, and Spanish find that the explicit metadiscourse expected in English seems unsophisticated and childish. Reader-responsible texts allow for a much more implicit organization of texts, where the reader is considered an intelligent being "to whom very little needs to be explained" (Mauranen, 1993, as cited in Fernández Polo, 1999, p. 281). The author of a book for translators is quoted as saying, "Spanish frequently prefers to leave much to the intelligence and to the imagination" (Vázquez Ayora, 1977, as cited in Fernández Polo, 1999, p. 281). One Spanish scientist laments that to adopt an explicit style would be like writing for "bobos" (St. John, 1987, p. 119). The reader in these Asian and European languages is expected to make a greater investment of effort in following the writer's text.

It must be pointed out that scientific writing conventions differ between disciplines, and talking in broad generalizations about "the scientific article" can be somewhat misleading. There has been much research that characterizes the rhetoric, generic structure, hedging, and boosting of evaluative statements and the conventions of evidence among different disciplines (see Hyland 2009, among many others). The advent of corpus linguistics has allowed for detailed mining of lexicon and collocation in research articles (Biber, Conrad, & Cortes, 2004; Byrd & Coxhead, 2010; Cortes, 2010; Hyland, 2008).

For the scientist writing in English as a second language, it is often necessary to violate the sensibilities and conventions of their first language to comply with textual demands of English science writing. Linguistic and cultural differences of scholars' first language and periphery location have led some applied linguists to call for greater recognition and respect for such variation by journal reviewers and editors (Benesch, 2001; Pennycook, 1994). International scholars are called upon to resist the homogenization of scholarly writing, which Heidi Byrnes termed "the virtual effacement of the L1 [first language] abilities and predispositions" (Byrnes, 2002, p. 42). It is clearly important to comply with the textual conventions of English scientific writing. This is pointed out by Robert Kaplan (2001): "Good scientists who cannot write English to meet the standards of journal editors are deprived of the opportunity to have their views and contributions disseminated through the global information networks" (p. 18). By being excluded, "their contributions ... are lost to science" (p. 18).

3.7 SUMMARY

This chapter provides a comprehensive review of the research regarding publishing in English on the part of scholars located outside Anglophone center countries and who write in English as a second language. It has been demonstrated that there is much publishing that international scientists engage in that is simply not counted. Their work is not included in international databases, such as the Web of Science, and consequently those articles are not even considered part of the scholarly output of a nation. To be included in the journals that receive international recognition, elaborate networks of collaboration have been found to be very useful. In some instances, international collaboration between a scholar located in a peripheral country and one located in a core country may be unequal or disproportionate. On the other hand, access to international publishing is greatly enhanced through the social practices and collegial networks that surround the scholarly text. The expectations of journal reviewers and editors may, in some cases, prejudice them against peripheral scholars, but this does not always occur, and some journals and editors actively welcome and nurture international submissions. Rhetorical and linguistic features in a manuscript may make the scholarly content more difficult to discern, and those features may emanate not from errors per se, but differences in linguistic and cultural conventions of scientific writing. To conclude this chapter, it seems clear that compliance with the conventions of scientific writing lessen the obstacles a peripheral scholar faces in negotiating a manuscript into print. Further, establishing strong networks of formal and informal collaboration with the dominant members of the discipline who are successful in publishing their work is greatly desirable. Finally, national and international policies to give credit to scholars' work that is published in languages other than English and in smaller journals will ultimately benefit all of science.

4 Methodology: Researching Spanish Speaking Scientists in Mexico

4.1 Introduction

The pervasiveness of English in scientific knowledge production is well documented, as demonstrated in the synthesis of research presented in the previous chapter. The research also demonstrates that being a non-native speaker of English, and being physically situated outside of the center English-speaking countries, presents extra difficulties for scientists who seek to publish their research in English-medium journals. However, to date, there are several avenues of research that have not yet been undertaken regarding this population. Specifically, there has been little research that attempts to quantify the difficulty of writing science in a second language in measurable terms. Further, there has not been a systematic attempt to distinguish subgroups within this larger category of English-as-a-second-language scientist. The educational trajectory that professional scientists located in the periphery have travelled has not been correlated to their experiences in publishing. Finally, the role of language and literacy within that educational trajectory has not been carefully examined. To address some of these questions, a three-year study was undertaken. The results of this study are reported in the coming chapters. The aim of this chapter is to outline the ways this study was conducted.

The research agenda for this study utilized both diachronic and synchronic approaches. With a view to changes over time, we were particularly interested in learning trajectories. Two research questions guided the diachronic aspect of the study:

1. How do scientists learn to write research articles for publication in English as a second language?
2. What are the best ways to teach second language scientists to write research articles in English?

To understand the synchronic aspects of scientists currently working in the periphery, two other questions guided the research:

1. To what degree is the process of scientific writing different in the author's first (L1) and second language (L2)?
2. Are there systematic differences in the process of writing for different second language scientists?

Investigating these questions involved 148 scientists (38.4% of those surveyed) in two institutions of higher education in Mexico, a Spanish speaking country. The overall research design involved a mixed method approach that combined quantitative and qualitative methods. Quantitative methods were used to explore the synchronic aspects of this study and qualitative methods to explore the diachronic aspects.

We begin this chapter with a brief introduction to the state of science in Mexico, and continue with a description of the sites, participants, methods, and analytical procedures. The findings from this methodology are presented in the next four chapters of this volume. Chapter 5 presents the results of the quantitative study. Chapter 6 addresses the narratives and understandings of 16 scientists who participated in the qualitative study. Chapter 7 details the qualitative findings in categories and groups, allowing us to present trends. Chapter 8 synthesizes these findings in the context of the research conducted around the world concerning writing science in a second language.

4.2 Science in Mexico

Mexico is a developing country that has limited bibliometric presence in terms of scientific production. Currently, Mexican scientists publish just under 7,000 articles a year in the journals counted by Web of Science. This number represents a steady increase in productivity over the past 25 years (Figure 4.1) and indicates the increasing importance of publishing in journals that are counted in such metrics. However, to contextualize this publishing record, figures show that a typical, top-tier American university publishes between 4,500 and 6,000 articles annually from that one institution (Mervis, 2010). Despite the increasing number of Mexican scientific articles, the country's participation in the world production of scientific articles is still very small, at less than one percent (Figure 4.2).

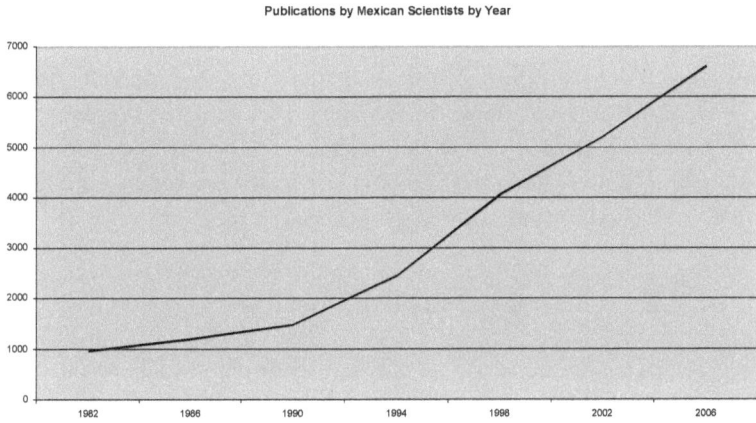

Figure 4.1 Number of scientific articles published by Mexican scientists each year from 1982 to 2006. Source: CONACYT, 2009.

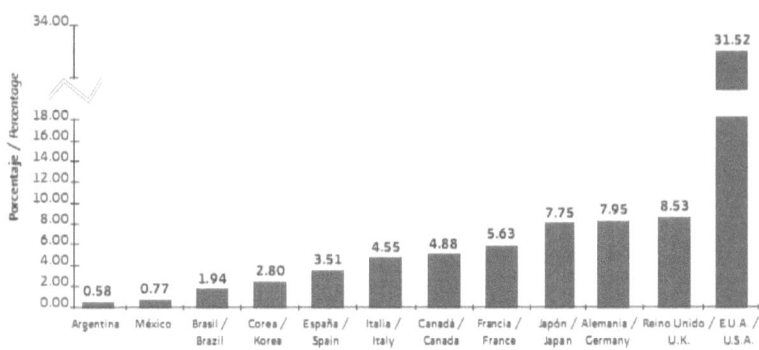

Figure 4.2 Percentage of countries' share of published articles in 2007. Source: CONACYT, 2009.

There is increasing emphasis placed on publishing scientific work in the journals that are included in international bibliometric measures. For the scientists themselves, salary and merit pay incentives are tied to publication, with different weight assigned to publishing in peer-reviewed versus non-peer-reviewed journals, and national versus international journals. Additionally, universities are funded on a formula that takes account of publishing productivity, so they in turn place increasing pressure on scientists to participate in the international research arena. Finally, the country itself belongs to international organizations such as the Organisation for Economic Cooperation and Development (OECD), the Organization of American States (OAS), and the World Bank, where Mexico is evaluated on many criteria including scientific and technological knowledge development. Performing well on such indicators is important not only for national pride, but access to international resources.

4.3 Site Description

The study was conducted at two public institutions of higher education in Mexico, a university and a research institution. Both institutions are located in the city of Ensenada (population 400,000), which sits beside the Pacific Ocean, 62 miles south of the U.S. border. In addition to these two public institutions, there are three private universities in the city that offer degrees at the undergraduate and graduate levels. Public institutions were selected for this study because, with only one or two exceptions, they have a higher status within Mexico than private ones. Competition for entry into programs of study is greater, and often the remuneration policies to faculty are more generous in public institutions. In other words, the best and the brightest gravitate there.

The first university research site is the Universidad Autónoma de Baja California (UABC). This is a public, state-wide university with three large campuses and several specialized satellite campuses spread across the state. The university is consistently rated as among the top five universities in the country. Enrollment is approximately 40,000 students per annum state-wide, of whom less than 5% are graduate students; however, there are intensive efforts in recent years to increase the graduate level offerings. Faculties of medicine, law, business administration, engineering, education, social sciences, and the humanities are located in campuses around the state.

The three natural science faculties, where this research was conducted, are located in the city of Ensenada. The Faculty of Science is comprised of biology, physics, mathematics, and computer science. The separate Faculty of Marine Science specializes in oceanology, environmental science, and biotechnology. The Institute for Oceanological Research offers graduate programs at the master's and doctoral levels. These natural science faculties enroll only 2% of the student body, but the university's international reputation for marine sciences attracts students from throughout the Spanish-speaking world. The teaching staff in these natural science faculties are the most productive among all the faculties at all the campuses in terms of research papers published, collaborative research investigations, and sabbaticals abroad.

The other research site of this study was the Center for Scientific Research and Higher Education of Ensenada (CICESE). It was created 35 years ago by the Mexican federal government as part of an effort to decentralize scientific research beyond the capitol of Mexico City and modernize the country. As an institution, the first of its three objectives is to "generate scientific knowledge through research projects." The second objective is to "form human resources through masters and doctoral programs." Third, the institute strives to "strengthen the link with public, private, and social sectors through research and development projects." It is comprised of four divisions: Earth Sciences (geology, seismology, and applied geophysics), Applied Physics (optics, computer science, and electronics and telecommunications), Oceanology (physical oceanography, biological oceanography, ecology, and aquaculture), and Experimental and Applied Biology (biotechnology, experimental microbiology, and biology of conservation). As a research institute, it offers no undergraduate programs. At the graduate level, 16 terminal degrees are offered, and an average of 80 masters and 20 doctoral degrees are completed each year. It is one of only ten independent research centers in the natural sciences in the country, and it produces the highest number of articles among those research centers (CONACYT, 2009).

4.4 Participant Population

The vast majority of the scientists at both these public institutions are Mexican nationals who speak Spanish as their first language. In addition, there are a handful of scientists who are Russian by birth, and who immigrated to Mexico approximately 20 years earlier as support for scientific endeavors were reduced with the collapse of the Soviet Union. There are a few Americans based at the research institute, and a few other immigrants, including scientists from Germany, Poland, and France. Only scientists who were Mexican and spoke Spanish as their first language participated in the current study. Scientists at the university who were based in one of the three natural science faculties were involved; in other words, none of the social sciences, humanities, or engineering faculties participated in this study. All the scientists at the research institute were involved. These scientists held a full range of scientific/academic positions, from those designated as "technicians" to the equivalent of "full professor/scientist." Consequently, the proportion of their time allocated solely to scientific research varied, and might also include teaching, administrative, and technical responsibilities.

The policies concerning faculty activities are different at the two institutions. University policies require that faculty split their time among four activities: teaching, research, administrative service, and student advising. There is no cap on the number of teaching hours expected, and while 4-16 hours per semester is typical, it may reach as high as 24 hours within a 40-hour week. In 2012, the National Organization of Researchers raised the minimum number of hours that public institutions must provide to the most productive researchers in order for them to conduct research: from 16 hours a week to 20 hours in a 40-hour workweek. All full-time faculty at the university are required to perform student advising activities. This involves guiding as many as 40 students each semester in their course selections, verifying their programs of study and other administrative roles. Remuneration and merit-pay policies take into account all four activities, although the greatest value is accorded to publishing research in international, English-medium journals. At the research institute, teaching rarely involves more than 4-6 hours per week, per semester, and all courses are at the graduate level. Informal policy expects researchers to publish at least one peer reviewed paper per year in a nationally or internationally indexed journal, and their productivity is periodically reviewed by department committees.

Statistics concerning researchers' publishing activities report that, on average, nationally, Mexican researchers who are members of the National Organization of Researchers published .72 papers in journals of the Science Citation Index, i.e. less than one paper per person, during 2009. At the sites for the current study, the total number of scientific articles published and citations those articles accrue is almost double at the research institute compared to the university (CONACYT, 2009); however, specific proportional comparisons between the university science faculties and the research institute are not available. For the purpose of our study, all participants indicated that they were involved in research, and they sought to publish their work. They reported that they were employed anywhere from six months to 34 years with their respective institution.

4.5 METHOD

This study addresses questions of both a synchronic and diachronic nature. To accomplish this, two separate studies were conducted in which a qualitative methodology followed from the initial quantitative methodological study. The methods and analytical approach used for each of these studies are addressed below.

4.5.1 Quantifying and Specifying the Difficulties of Writing a Research Article for Publication in a Second Language: Method and Analytical Approach

This portion of the study with Mexican scientists addressed synchronic issues surrounding writing scientific articles in English. Specifically, we wanted to know the degree to which the process of L2 science writing for publication was different from L1 science writing (if at all), and whether there are systematic differences in the process of writing for different second language scientists. To explore these questions, a quantitative survey methodology was used. A 16-item survey instrument was developed and distributed to the participants for their responses.

To understand the self-reported differences between Spanish (L1) and English (L2) scientific research writing, a differential *D-score* approach was used. As a starting point, the concept of difference in writing was conceptually analyzed. Since the aim of the study was to explore the potential differences in relation to the possible added burden that L2 science writing might involve, the concept of burden in writing was differentiated in relation to the self-evaluation of difficulty, anxiety, and satisfaction. Using the *D-score* approach, the same questions concerning difficulty, anxiety, and satisfaction were asked in relation to L1 (Spanish) and L2 (English) science writing. Conceptually, the difference between the scores for the same item on L1 and L2 writing represents the added burden of L2 writing (if such a burden actually exists). The six items (three in relation to Spanish; three in relation to English) were set out as Likert scales ranging from 1 to 7 (Appendix A presents the full version of the questionnaire). The specific questions and their format for this section of the survey were as follows:

1. On the scale below, rank the degree to which you find it *easy or difficult* to write a scientific article in Spanish for a major journal.
    Very Easy | 1 2 3 4 5 6 7 | Very Difficult

2. On the scale below, rank the degree to which you find it *easy or difficult* to write a scientific article in English for a major journal.
    Very Easy | 1 2 3 4 5 6 7 | Very Difficult

3. On the scale below, rank the degree to which you are *satisfied or dissatisfied* that your writing in Spanish conveys the scientific research that you have conducted.
    Very Satisfied | 1 2 3 4 5 6 7 | Very Dissatisfied

4. On the scale below, rank the degree to which you are *satisfied or dissatisfied* that your writing in English conveys the scientific research that you have conducted.
    Very Satisfied | 1 2 3 4 5 6 7 | Very Dissatisfied

5. On the scale below, rate the degree to which writing a scientific article in Spanish for a major journal causes you to feel *anxiety*.
    Not at all Anxious | 1 2 3 4 5 6 7 | Very Anxious

6. On the scale below, rate the degree to which writing a scientific article in English for a major journal causes you to feel *anxiety*.
    Not at all Anxious | 1 2 3 4 5 6 7 | Very Anxious

Other items on the survey instrument were intended to obtain data on other aspects of the scientists' writing experiences. Relevant questions asked the scientists to indicate:

- The percentage of their scientific publishing performed in English;
- The degree to which English is a barrier to their publishing efforts;

- The degree to which they believe journal editors are biased against authors with non-Anglo-Saxon names;
- The type of writing processes they used; and
- Basic biographical data concerning position, institution, and years of employment.

No one's personal name or identifying characteristics were sought.

The survey instrument was produced in Spanish. It was distributed both in paper form and electronically through Survey Monkey, in several iterations, in the fall of 2007. Distribution at the research institute covered the scientists at all four divisions: Earth Sciences, Applied Physics, Oceonology, and Experimental and Applied Biology. The survey questionnaires were distributed to all full-time research faculty, 216 in number. Seventy questionnaires were returned. At the university, teachers and researchers at only the three natural sciences faculties of Sciences, Marine Sciences, and the Institute for Oceanological Research were included. A total of 169 questionnaires were distributed and 68 were returned. An additional 10 surveys were returned without indicating at which institution the participant was based. In total, 148 surveys were returned, constituting a 38.4% response rate from both institutions. The number of responses was virtually evenly split between the two institutions. This large representative participation on the part of the scientists at institutions in Mexico provides a strong basis on which to document the results and conclusions.

4.5.2 Developing Scientific Writing Expertise: Method and Analytical Approach

An important element of our study was to understand the diachronic development of scientific writing expertise. To explore these aspects of scientific writing, a qualitative approach is appropriate. Upon completion of the quantitative study outlined above, a qualitative study was undertaken. This second study was based and designed in relation to the findings of the quantitative analysis. Specifically, one of the important findings was that the population of scientists explored in this study divided into four separate groups according to seniority (junior/senior) and institution (research institute/teaching university).

As the results from the quantitative analysis indicated that there are four distinct groups within our population, we sought participants who were representative of those groups: senior researchers from the research institute, junior researchers from the research institute, senior researchers from the university, and junior researchers from the university. To identify these researchers, an email invitation was sent to *all* the same scientists who had been contacted for the original survey. We asked the researchers to send us their curriculum vitae, and assured them that their anonymity would be protected. In total, 48 did so: 28 were from the research institute, and 20 were from the university. These vitae were then divided into junior and senior faculty in the two institutions using six years of academic activity within the institution as the demarcation criterion. Six years was chosen for demarcation because this is when tenure would be obtained in a North American context. Four groups of faculty with curriculum vitae were created: university junior faculty; university senior faculty; research institute junior faculty; and research institute senior faculty.

Following a review of the curriculum vitae of potential participants in the four groups, four scientists from each group were chosen to be interviewed. Interviewees were chosen for their representativeness of their respective group and for their publishing activities presented on their vitae. Four scientists from each of these groups were then interviewed for a total of 16 interviews conducted. Upon being selected for an interview, each scientist was asked to provide a copy of three articles they had written that "they were proud of." Prior to the face-to-face, we read those articles and reviewed the scientists' curriculum vitae.

The interviews were held in the scientist's office to provide the highest level of professional courtesy to the participant, and were conducted in Spanish. The interviews followed a semi-structured format, covering seven major themes in their professional lives. The interview protocol was as follows:

Question 1: Educational (Scientific Literacy) Background
Think back over your educational and professional career. How did you learn to write a research article in English? Please consider formal educational experiences as well as informal interactions with colleagues or other people. Seriously consider all the settings and people who have helped you to learn to write a research article in English.

Question 2: Personal Perceptions of Strengths and Difficulties
Think back over your publication career in English. What would you say are your greatest strengths and difficulties as a writer of scientific

articles in English? Please provide specific examples for each of the points you make. How do you overcome the difficulties that you have specified?

Question 3: L2 Scientific Writing Process

Let's look now at the three articles you chose to give me prior to this interview. I am interested in the way these specific articles were written. Could you try and describe for me all the stages of writing these articles? I am particularly interested in the languages you used and how you worked with other people involved in this publication. Please be as specific as possible in describing all the stages of writing.

Question 4: Perceptions of Successful and Failed Writing Processes

Think for a moment about your experiences as a writer of scientific articles. Could you describe a writing experience that really worked very well for you? What was it about that process that was so successful? Could you now think of a writing experience that really went badly? What was it about that process that was so flawed? Please try to be specific in your answers.

Question 5: Perceptions of Helpful Solutions

Let's say that I was awarded one million dollars to set up a language center designed to help you in your work as a publishing scientist. What would you like me to be able to help you with? What kind of help would be effective for you? What I am asking is your sense of what would be the most effective use of this money in facilitating your scientific publication.

Question 7: Open Discussion

Is there anything you would like to add to this interview? Are there any aspects of the writing process that you feel I have not addressed and you would like to discuss?

The interviews were tape recorded and subsequently transcribed and translated.

Analysis of the interviews was performed by successive readings in four topic areas. First, each interview was read to create a timeline of the scientists' educational trajectories and the accompanying literacy activities at each stage. Second, we looked closely for the scientists' description of their role in writing the three articles they had provided to us as examples of articles they were proud of. These were analyzed based on the their self-described role and process of writing, the collaborators involved and their roles, and their assessments of the efficacy of the collaborative efforts. Third, we extracted and categorized the scientists' comments about their strengths and weaknesses as writers of scientific articles in English. Fourth, we collected the scientists' recommendations about what types of assistance they believed would be helpful to facilitate their successful publishing in English. Finally, we took separate account of other noteworthy comments. After this analysis of each of the sixteen interviews, group data were determined, looking for trends, consistencies, and variety in the self-reported interview data.

4.6 SUMMARY

This study aims to contribute to the body of knowledge concerning writing scientific articles in English on the part of non-native speakers of English. To do so, both diachronic and synchronic perspectives were adopted concerning how scientists learned to write scientific articles in English and how they currently manage that activity. Initially, a questionnaire was distributed to all the scientists in the natural and applied sciences at two institutions of higher education in Mexico. The results of this quantitative study motivated an additional qualitative study. In this second study, 16 scientists were interviewed, and they each provided copies of their curriculum vitae. There were four scientists in each of four subgroups: junior and senior scientists at the research institute, and junior and senior scientists at a large public university. In addition to participating in an interview, these scientists provided an up-to-date curriculum vitae and copies of three articles they had written. The interviews were transcribed and analyzed for individual and group trends that addressed the study's research questions. The results of these analyses are presented in the next chapters.

5 The Quantification and Specification of the Difficulties of Writing a Research Article for Publication in a Second Language: Survey Report

5.1 INTRODUCTION

This chapter reports on the results of the survey of 148 active scientists from two institutions of higher education in Mexico. As described in the methodology chapter (Chapter 4), this study was designed to elicit quantitative data on the self perceptions of scientists who publish in English as a second language (ESL), and thus provide additional data on issues faced by this specific population in their professional writing. This data provided the backdrop for the second stage of this study (Chapter 6), which followed up these initial findings with an in-depth interview process and document analysis. The core question explored in this chapter addresses the relative burden placed upon second language scientists in relation to their L2 English science publication experiences and their perceptions of the difficulties they face. As described in the methodology chapter, a special aspect of the design of this study was the attempt to differentiate scientific writing in different languages and its influence on the self perception of scientific writers through the comparative analysis of four groups of scientists—junior and senior faculty at both a teaching university and a research institute. To isolate the second language writing factor from other issues of scientific competence (such as the actual quality of the scientific research), a decision was made to compare self perceptions in relation to first language (Spanish) and second language (English) scientific writing. The differences among the responses to the L1 questions, as compared to the L2 questions, provided a way to quantify the actual burden of writing scientific articles in a second language, and thus directly address the increased difficulty of scientific writing in a second language.

5.2 Research Questions

The quantitative survey component of this study was designed to answer the following research questions:

1. To what degree is writing a scientific article for publication in English as a second language more difficult than writing a scientific article in a first language?
2. Are these difficulties shared by researchers in different scientific environments and differing levels of seniority?
3. What are the specific difficulties in writing a scientific article in English reported by scientists who are writing in a second language?

5.3 Results

The first issue is the core question that addresses the relative burden placed upon second language scientists in relation to their L2 English writing for publication and their perceptions of the difficulties they face. As described in the methodology section, the concept of language writing burden was conceptualized as the differential response between self-rated perceptions of the relative difficulty, satisfaction, and anxiety between first and second language scientific writing for publication. Operationally, this was set out as a series of parallel questions with rating scales in relation to the difficulties, degrees of satisfaction, and levels of anxiety associated with scientific writing in both first and second languages (see questions 1-6 on the questionnaire). Accordingly, the first issue addressed within this results section is the descriptive data for rating responses for first (Spanish) and second language (English) scientific writing. Figure 5.1 summarizes the means and standard deviations for self perception ratings on questions of difficulty, satisfaction, and anxiety in first (Spanish) and second language (English) scientific writing for all participants.

(a)

Question	Mean	S.D.
Difficulty of Spanish	2.45	1.48
Difficulty of English	4.25	1.86
Dissatisfaction in Spanish	2.15	1.39
Dissatisfaction in English	3.02	1.98
Anxiety in Spanish	2.57	1.60
Anxiety in English	4.20	1.98

(b)

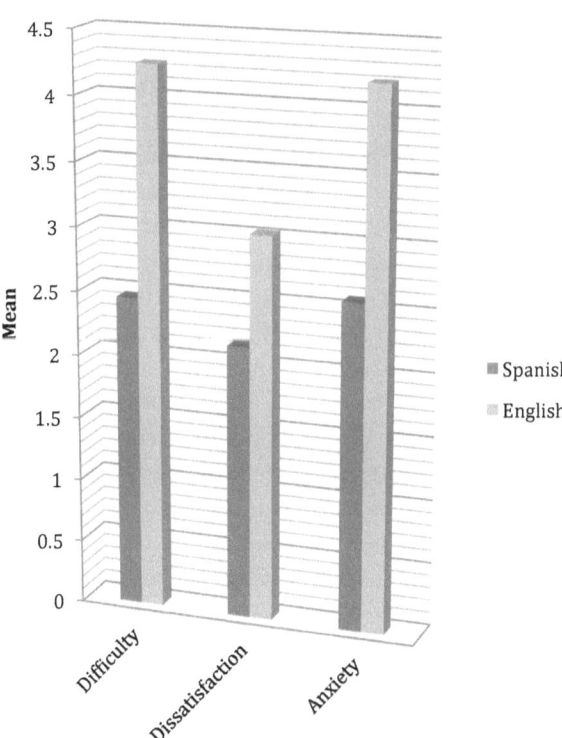

Figure 5.1 Means (figs. 5.1a & 5.1b) and standard deviations (fig. 5.1a) for self perception ratings on questions of difficulty, satisfaction, and anxiety in first language (Spanish) and second language (English) scientific writing (n=148)

As can be seen in Figure 5.1, for the scientists who participated in this survey, there are differences in their average ratings of the difficulty, degree of satisfaction, and levels of anxiety involved in writing a scientific article in English as a second language when compared to their first language scientific writing. The results indicate that, on average, writing an article always presents some level of difficulty, anxiety, and dissatisfaction with the final product. This is represented in the scientists' self-reports concerning producing a scientific article in Spanish, their L1. On a seven-point scale from 1 to 7, where higher numbers indicate greater difficulty, dissatisfaction, and anxiety, in all factors the mean is larger than 1, which indicates there is some difficulty, some dissatisfaction, and some anxiety associated with writing a scientific article even in their first language. Specifically, scientific writing in English as a second language (L2) was perceived, on average, to be 24.14% (1.69 points on a 7-point scale) more difficult than scientific writing in L1; the final written article in ESL was perceived to be 11.71% (-0.82 points on a 7-point scale) less satisfactory than a scientific article written in L1; and finally, ESL science writing was reported to induce 21.71% (1.52 points on a 7-point scale) more anxiety than writing in L1. This descriptive data suggests that ESL scientific writing is quantifiably more difficult, less satisfying, and generates more anxiety than scientific writing in a first language.

One of the issues of interest in the current study is the role of different educational settings for scientific activity on self perceptions of the burden placed upon second language scientists in relation to their professional L2 English writing. Within the current study, this was operationalized using a quasi-experimental design by collecting data from scientists in two settings—a teaching university and a research institute (see the "Site Description" section in the methodology chapter for a thorough description). These two settings represent the two common sites at which research is conducted around the world. A further assumption behind this design was that a range of factors—such as greater access to international colleagues, difference in time allocation, supervisory and service requirements, and publishing expectations–may differentiate the work of scientists in these two settings, and thereby influence the self reported responses of these participants. Accordingly, the descriptive data presented above can be further broken down into responses between these two institutions. Figure 5.2 summarizes the means and standard deviations for self perception ratings on questions of difficulty, satisfaction, and anxiety in first (Spanish) and second language (English) scientific writing (questions 1-6 of the questionnaire) for participants at the teaching university (TU) and the research institute (RI).

(a)

Question	Mean TU (n=73)	S.D.	Mean RI (n=75)	S.D.
Difficulty of Spanish	2.25	1.29	2.57	1.58
Difficulty of English	4.82	1.83	3.75	1.76
Dissatisfaction in Spanish	2.45	1.48	1.90	1.30
Dissatisfaction in English	3.89	2.09	2.19	1.47
Anxiety in Spanish	2.56	1.68	2.54	1.52
Anxiety in English	4.73	1.93	3.69	1.91

(b)

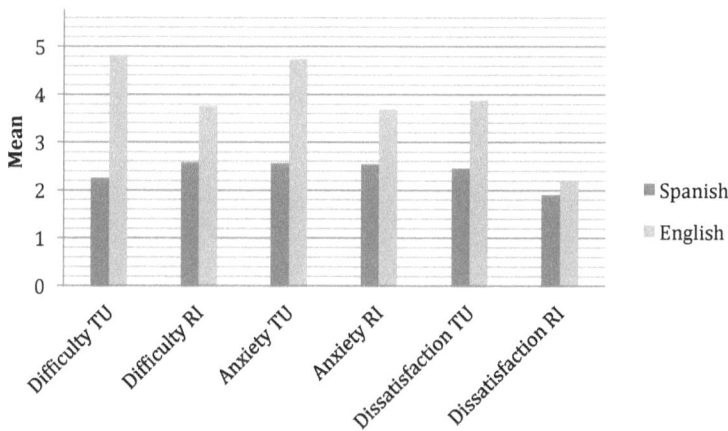

Figure 5.2 Means (figs. 5.2a & 5.2b) and standard deviations (fig. 5.2a) for self perception ratings on questions of difficulty, satisfaction, and anxiety in first language (Spanish) and second language (English) scientific writing for faculty at a teaching university (TU) and a research institute (RI).

As can be seen in Figure 5.2, the same trends in relation to the burden of ESL science writing can be found in both the teaching university and the research institute, with ESL writing being rated as more difficult and generating higher levels of anxiety and dissatisfaction. However, the difference in rating between English and Spanish seems to be more pronounced in the teaching university than the research institute. For all three ESL questions, faculty at the teaching university had higher ratings than their counterparts in the research institute. Faculty at the teaching university had greater difficulty in ESL writing, and this generated higher levels of anxiety and dissatisfaction. In addition, the relative burden of ESL science writing was different for the faculty from the two settings. On average for the teaching university faculty, scientific writing in ESL was perceived to be 36.71% (2.57 points) more difficult than scientific writing in L1; whereas, for the research institute faculty, this was perceived to be only 16.85% (1.18 points). For the teaching university faculty, scientific writing was rated as involving 31% (2.17 points) more anxiety than writing in L1; whereas, for the research institute faculty, this was rated as involving only a 16.42% (1.15 points) increase in anxiety. For the teaching university faculty, ESL science writing involved a 20.57% (1.44 points) increase in the degree of dissatisfaction with writing when compared to L1 science writing; whereas, for the research faculty, ESL science writing involved only a minor, 4.14% (0.29 points) decrease in satisfaction. These results suggest that while ESL science writing is a burden for both groups, it is more of a burden for the teaching university faculty than the research institute faculty.

To evaluate the above analysis of the descriptive data, a two-way MANOVA was calculated using language (Spanish/English) and institution (Teaching University/Research Institute) as the independent variables, and difficulty ratings, dissatisfaction ratings, and anxiety ratings as three dependent variables. Hotellings' T^2 multivariate generalization of the univariate t value was used. The MANOVA revealed a highly significant effect for language (Hotellings' T^2=30.392, $p<0.000$); a highly significant effect for institution (Hotellings' T^2=11.527, $p<0.000$); and a significant interaction between language and institution (Hotellings' T^2=4.296, $p<0.006$). To further explore these results, univariate F-tests were calculated to determine which variables contributed to the interaction and overall difference. Significant interactions for language and institution were found for all three dependent variables: difficulty, $F(1,253)=31.07$, $p<0.001$; dissatisfaction, $F(1,253)=20.84$, $p<0.05$; and anxiety, $F(1,253)=15.73$, $p<0.026$. The univariate F-tests were significant for language comparisons

(Spanish/English) on all three variables: difficulty, $F(1,253)=84.11$, $p<0.000$; dissatisfaction, $F(1,253)=18.58$, $p<0.000$; and anxiety, $F(1,253)=55.24$, $p<0.000$. In comparisons between institutions (Teaching University/Research Institute), the univariate F-tests were significant for two of the variables: dissatisfaction, $F(1,253)=80.82$, $p<0.000$ and anxiety, $F(1,253)=16.11$, $p<0.026$. But the univariante F-tests showed only a trend in relation to the third language variable: difficulty $F(1,253)=3.33$, $p<0.069$. In sum, the inferential statistics presented above support the position that ESL science writing is significantly different from L1 Spanish science writing, and that writing science in English as a second language is an additional burden for the second language scientist. In addition, the results support a difference between the two educational settings used in this study. The burden of ESL science writing is more pronounced for the faculty at the teaching university than the research institute.

To further explore the difficulties faced by ESL science writers, a series of questions within the questionnaire explored the ramifications of ESL writing on publication. The most direct question of this type (Question 9) asked participants to rate the degree to which writing in English was a significant barrier to publishing their research. On a seven-point scale (with 7 indicating writing as a very significant barrier to publication), faculty at the teaching university gave an average rating of 4.13 (S.D. 2.04), signifying that they considered ESL writing to be a barrier. The research institute faculty gave an average rating of 2.68 (S.D. 1.78), signifying that they considered ESL writing to be a low barrier. Twenty-three of 79 of scientists from the research institute—almost one third—reported that English presented no barrier. However, it should be noted that 55% of all respondents gave a rating of above 3.5, signifying that for over half of the respondents, ESL writing was considered a problem in relation to scientific publication. Twenty percent of the respondents gave a rating of 6 or 7, signifying that this group considered ESL writing to be a very significant barrier to publishing scientific results. In addition, 66.7% of respondents specified that they had received comments about language from editors or reviewers during the peer review process of article submission with 14.11% of respondents specifying that their English writing was criticized on 70% to 100% of all the papers they had submitted for publication. Fifty-four percent of respondents stated that they had been critiqued only one to three times. The picture that emerges from this data is that for the majority of the second language scientists investigated in this survey, ESL writing for publication represented a barrier to publishing scientific results. This barrier was not, however, uniform, and is not an accurate representation of all participants. For some participants—particularly those from the research institute—ESL writing was not perceived as a significant barrier.

An additional aspect of this study concerns the types of difficulties that ESL scientists report in relation to their writing. One question in the questionnaire (Question 13) asked for a brief description of the difficulties faced by ESL science writers. Table 5.1 provides a summary of the respondents' answers. As can be seen in Table 5.1, 52.6% of the responses from the ESL scientists specified structural linguistic issues in English as a cause for difficulties in writing. The most frequently stated difficulty (31.1%) addressed linguistic elements at the level of the sentence, while 17.2% of the responses addressed the process of writing and the inherent problems involved in actually writing a science paper in English. Together, the language factors (linguistic form, genre, style, and writing process) consisted of 69.8% of all the stated difficulties in writing a science article in English.

Table 5.1 Summary of written statements made by participants concerning the difficulties they face in writing a scientific article in English (n=93)

Category	Frequency	Percentage
Language Form	29	31.1
Genre form and style	20	21.5
Writing processes	16	17.2
Scientific Content	14	15.0
Editorial bias	12	12.9
Institutional Factors	2	2.1

Additional difficulties were presented by the participants, including issues with scientific content, institutional barriers, and editorial bias. This last issue of editorial bias was addressed directly in another question on the questionnaire (Question 12). Respondents were asked to rate on a seven-point scale the degree to which they found editors and reviewers to be fair in their review of manuscripts written by scientists with Mexican names or from Mexican institutions. Eighty-three, or 64.9%, of respondents gave a rating of 4 or above, signifying that the majority of the respondents found editors' and reviewers' reviews to be fair to very fair. Of this group, 35.2% gave a rating of 6 or 7, signifying that they found the review process to be very fair, and 19.5% of respondents gave a rating of 1 or 2, signifying that they felt that the review process was very unfair to unfair. While the majority of ESL science writers feel they are treated fairly, there is a minority who feel they are very unfairly treated in the review process because of their second language writing, non-Anglo-Saxon names, and/or non-Anglo institutional affiliation.

The second language scientists in this survey were also asked about the overall writing strategy used in writing a research article in English. Several different options for addressing this task were presented to the participants, in addition to the option of specifying an individual process (see Question 10). Table 5.2 presents a summary of responses, organized in relation to the two educational settings, the teaching university and research institute.

Table 5.2 Percentage and frequency of writing strategy options for faculty at the teaching university and the research institute.

Writing Strategy	Percentage and Frequency Teaching University	Percentage and Frequency Research Institute
Trust my own writing	27.1% (16)	49.2% (30)
Write collaboratively	16.9% (10)	24.6% (15)
Share with Native Speaker of English	23.7% (14)	22.9% (14)
Translate	30.5% (18)	1.6% (1)
Translate and Share with Native Speaker of English	1. 7% (1)	0
Use Translator Software	0	1.6% (1)

As can be seen in Table 5.2, there are differences between the ways scientists in the two educational settings conduct the writing of a scientific paper in English. The biggest difference relates to the use of translation as a writing strategy by the 30.5% of the faculty in the teaching university. This strategy of writing was used by only one scientist in the research institute. Within the research institute, the majority of faculty (73.8%) either trusted their own English writing abilities or worked collaboratively with other scientists without approaching a native speaker of English; for the teaching university faculty, 44.5% also either trusted their own writing or worked collaboratively with other ESL scientists. Teaching university faculty rely more on translation and interaction with native English speakers than the research institute faculty. These results once again suggest that there are differences among second language scientists, and that they cannot be addressed as a single, homogenous group.

The final issue to be addressed within this results section is the further specification of differences among ESL science writers. As described in the methodology section, part of the design was the collection of data from junior as well as senior faculty at the two educational settings. The idea behind this decision was that senior faculty may have greater experience as publishing scientists, while junior faculty may have had much greater exposure to the English language and instruction. Figures 5.3a and 5.3b present means and standard deviations for junior and senior faculty at the teaching university and the research institute.

An analysis of Figures 5.3 and 5.4, and Table 5.3 reveals several trends of interest. First, as reported in relation to the overall results for the two settings, the distance between L1 and L2 responses are more pronounced for all the teaching university faculty than for the research institute faculty. What is interesting in the current analysis is that this distinction is even more pronounced when the two ends of the spectrum are considered. Senior faculty at the research institute report only moderate differences between ESL science writing and L1 Spanish writing. For the senior faculty members at the research institute, ESL science writing is 15.5% (1.09 points) more difficult, involves an increase of 14.28% (1.00 point) in anxiety, and a negligible 2.85% (0.2 points) decrease in satisfaction. In contrast, the junior faculty at the teaching university report a large difference between their ESL science writing and L1 Spanish writing. For the junior faculty at the teaching university, ESL writing is 40.71% (2.85 points) more difficult, involves an increase of 39.57% (2.77 points) in anxiety, and leads to a 19.71% (1.38 points) decrease in satisfaction.

(a)

Question	Mean Senior (n= 45) TU	S.D.	Mean Junior (n=13) TU	S.D.
Difficulty of Spanish	2.29	1.25	2.15	1.57
Difficulty of English	4.74	1.81	5.00	2.08
Dissatisfaction in Spanish	2.56	1.55	2.08	1.32
Dissatisfaction in English	3.93	2.14	3.46	1.94
Anxiety in Spanish	2.65	1.68	2.46	1.81
Anxiety in English	4.53	1.95	5.23	1.83
English as a Barrier	4.11	1.91	3.92	2.53

(b)

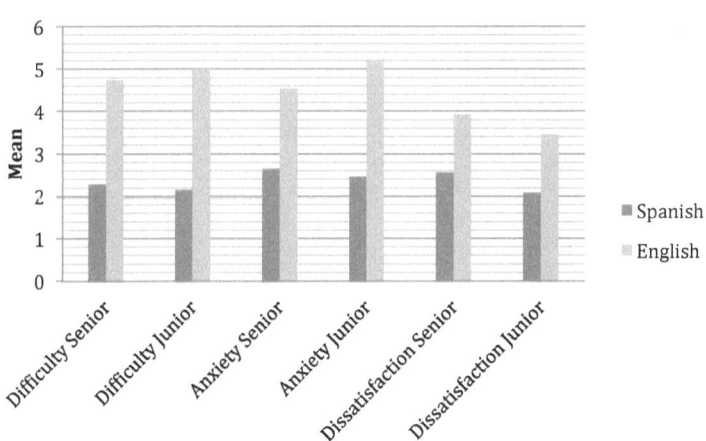

Figure 5.3 Means (figs. 5.3a & 5.3b) and standard deviations (fig. 5.3a) for self perception ratings on questions of difficulty, satisfaction, and anxiety in first language (Spanish) and second language (English) scientific writing for junior and senior faculty at a teaching university.

Senior faculty at the teaching university also seem to face difficulties in relation to their ESL science writing. They report that ESL writing is 35% (2.45 points) more difficult, involves an increase of 26.85% (1.88 points) in anxiety, and leads to a 19.57% (1.37 points) decrease in satisfaction. These results mirror the previous overall results for these two educational settings, supporting the position that these two educational settings involve different perceptions of the difficulties of ESL science writing.

(a)

Question	Mean Senior (n=54) RI	S.D.	Mean Junior (n=13) RI	S.D.
Difficulty of Spanish	2.52	1.57	2.77	1.69
Difficulty of English	3.61	1.77	4.15	1.68
Dissatisfaction in Spanish	1.87	1.41	2.00	0.82
Dissatisfaction in English	2.07	1.38	2.69	1.79
Anxiety in Spanish	2.33	1.36	3.23	1.88
Anxiety in English	3.33	1.79	4.92	1.75
English as a Barrier	2.44	1.69	3.38	1.85

(b)

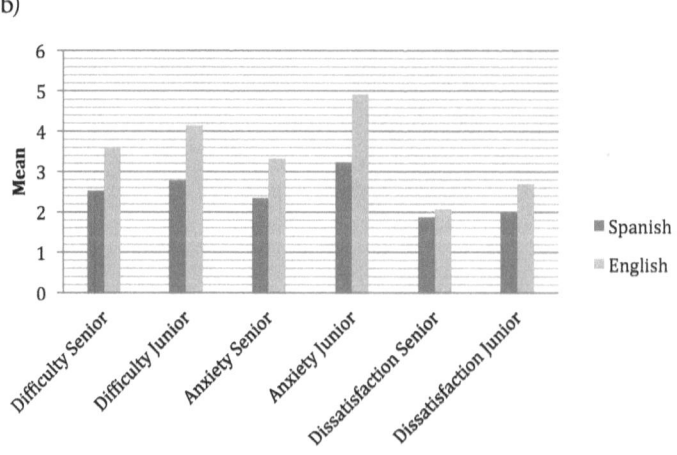

Figure 5.4 Means (figs 5.4a & 5.4b) and standard deviations (fig. 5.4a) for self perception ratings on questions of difficulty, satisfaction, and anxiety in first language (Spanish) and second language (English) scientific writing for junior and senior faculty at a research institute (RI).

Table 5.3 Summary of degree of difference for self reported ratings on questions of difficulty, satisfaction, and anxiety in first language (Spanish) and second language (English) scientific writing for junior and senior faculty at a teaching university (TU) and a research institute (RI).

Question Comparison	Percentage and Point Difference for Senior RI	Percentage and Point Difference for Junior RI	Percentage and Point Difference for Senior TU	Percentage and Point Difference for Junior TU
Degree of Increased Difficulty of ESL Writing	15.50% 1.09	19.71% 1.38	35.00% 2.45	40.71% 2.85
Degree of Increased Dissatisfaction of ESL Writing	2.85% 0.2	9.85% 0.69	19.57% 1.37	19.71% 1.38
Degree of Increased Anxiety in ESL Writing	14.28% 1.00	24.14% 1.69	26.85% 1.88	39.57% 2.77

An analysis of Figures 5.3 and 5.4, and Table 5.3 reveals several trends of interest. First, as reported in relation to the overall results for the two settings, the distance between L1 and L2 responses are more pronounced for all the teaching university faculty than for the research institute faculty. What is interesting in the current analysis is that this distinction is even more pronounced when the two ends of the spectrum are considered. Senior faculty at the research institute report only moderate differences between ESL science writing and L1 Spanish writing. For the senior faculty members at the research institute, ESL science writing is 15.5% (1.09 points) more difficult, involves an increase of 14.28% (1.00 point) in anxiety, and a negligible 2.85% (0.2 points) decrease in satisfaction. In contrast, the junior faculty at the teaching university report a large difference between their ESL science writing and L1 Spanish writing. For the junior faculty at the teaching university, ESL writing is 40.71% (2.85 points) more difficult, involves an increase of 39.57% (2.77 points) in anxiety, and leads to a 19.71% (1.38 points) decrease in satisfaction. Senior faculty at the teaching university also seem to face difficulties in relation to their ESL science writing. They report that ESL writing is 35% (2.45 points) more difficult, involves an increase of 26.85% (1.88 points) in anxiety, and leads to a 19.57% (1.37 points) decrease in satisfaction. These results mirror the previous overall results for these two educational settings, supporting the position that these two educational settings involve different perceptions of the difficulties of ESL science writing.

Another trend is that junior faculty in both the teaching university and the research institute are more anxious about their ESL science writing than their senior counterparts. For both groups of junior faculty, ESL science writing involves significant levels of anxiety. Junior faculty at the research institute report a 24.14% (1.69 points) increase in their anxiety levels when writing in English; junior faculty at the teaching university report a 39.57% increase in anxiety. An overall analysis of Table 5.3 shows that as we look at the four groups of faculty members, we see a progression in the relative burden on the faculty as a result of the ESL writing, with senior research institute faculty at one end of the spectrum, and junior teaching faculty at the other.

5.4 The Quantification and Specification of the Difficulties of Writing a Research Article in a Second Language: A Summary

The results presented in Section 5.3 above support the position that writing a scientific research article in a second language is perceived by active publishing scientists to be significantly different from writing a research article in their first language. As reported above, for the total population of second language scientists studied in this survey, this extra burden of ESL science writing consists of a 24.4% increase in difficulty, a 21.71% increase in anxiety, and 11.71% decrease in satisfaction within the final outcome. 55% of respondents felt that ESL writing was a barrier to scientific publication. Overall, the scientists who participated in this survey situated 69.8% of their difficulties in relation to the language factors of linguistic form, genre style, and the writing process, and the majority of respondents did not feel they were being unfairly evaluated during the review process. These results posit that from the perspective of active publishing scientists, writing scientific articles in a second language does indeed pose a problem that needs to be addressed.

However, the data presented here also suggests that the population of second language scientists is not homogenous, and different groupings can be found in the data we collected. The major division is between the two educational settings of the scientists. Faculty who work in the research institute report that English is less of a barrier to their scientific publication. While there are still statistically significant differences in their reported rates on the ESL and the L1 Spanish writing questions, the distances are smaller than for their counterparts in the teaching university. English writing for junior faculty is 19.71% more difficult, involves 24.14% more anxiety, and the results are 9.85% less satisfying than writing in their L1. For senior faculty at the research institute, their ratings for first and second language writing are, on average, smaller (English writing is 15.5% more difficult, involves 14.28% more anxiety, and the results are only 2.85% less satisfying).

For the faculty at the teaching university, English poses a much greater barrier to publication. The teaching university faculty rated their English science writing as 36.17% more difficult than writing in Spanish, as generating 31% more anxiety, and a has 20.57% decrease in satisfaction with the final product. For the junior faculty in the teaching university, these results were even more pronounced. Junior faculty at the teaching university reported that English science writing was 40.71% more difficult than L1 Spanish science writing, involved a 39.57% increase in anxiety, and led to a 19.71% decrease in satisfaction with the final outcome. For both the junior and senior faculty at the teaching university, ESL science writing was considered to be a major barrier to the publication of their research. As discussed in relation to Table 5.3, a clear progressive continuum of difficulty can be seen for the four groups studied in this survey.

The results in relation to the specification of where the difficulties of ESL science writing are situated suggest that we need to consider both the educational context and the degree of seniority of the scientists. Senior faculty at a research institute may need less help with their ESL science writing than other groups but still report on degrees of increased difficulty and anxiety. The self-reported increased difficulty and anxiety levels of ESL science writing for junior faculty at the research institute suggest that they would benefit from an educational intervention designed to enhance their scientific publication. The junior and senior faculty at the teaching university clearly require help. The junior faculty are especially worrisome in this context. The distance between their L1 Spanish science writing and ESL science writing is very large. Also, it should be remembered that junior faculty face much greater pressure in relation to publication than more senior faculty with tenure and that providing resources for junior faculty can be seen as a long term investment where early intervention may have career-length benefits.

5.5 Final Comments

The results of the study specify that language is indeed a quantifiable factor that contributes to the difficulty, dissatisfaction, and anxiety in writing research articles in English for the scientists who participated in this study, and that for the majority of the scientists explored in this survey, some form of educational intervention is required. The data presented in this chapter presents a quantitative view of this population and has helped in quantifying the degree of the burden involved in ESL science writing and distinguishing between different types of scientists and their relative needs. The next chapter continues this direction of research by providing qualitative data to allow a more in-depth understanding of these scientists and their ESL writing processes. Specifically, the aim of the next chapter is to present data concerning the educational trajectories, successful and unsuccessful ESL science writing experiences, and perceptions of solutions that would help them to move forward in their ESL science writing.

6 Developing Scientific Writing Expertise: Qualitative Individual Data

6.1 Introduction

This chapter reports on the results of the interview and document analysis (curriculum vitae and published papers) studies of a subset of the L2 scientists that formed the basis for the results reported in the last chapter. The aim of the chapter is to specify the processes by which the writing of research articles was learned by L2 science writers, the specific processes used by L2 scientists in writing their research articles, and their self-perceived strengths and weaknesses as science writers in a second language. The overall aim of this analysis is to describe those events and writing experiences that were useful for the enhancement of their English science writing. Further, the comments of L2 scientists are considered regarding the kinds of assistance, if any, they believe would be useful to improve their English science writing at this time in their careers. Accordingly, this chapter reports on the results of a qualitative investigation that had a double purpose: (1) to understand the process and difficulties of writing a scientific research article on the part of scientists working in English as a second language; and (2) to know in what ways scientists writing in English as a second language, and who live outside the U.S. and U.K. centers, learned to write scientific articles.

Our investigation consisted of 16 in-depth interviews conducted in Spanish with researchers at the two institutions involved in this study: a research university and a research institution. This study follows the results of the quantitative investigation presented in Chapter 5 that demonstrated there are statistically significant differences in the populations of the two institutions and that there were differences between senior and junior researchers.[2] Accordingly, the overall design of our study involved qualitatively investigating by means of interviews and document analysis four scientists from each of the four categories identified quantitatively: junior scientists at the research institution, junior scientists at the research university, senior scientists at the research institution, and senior scientists at the research university.

To facilitate a clear understanding of the positions and understandings presented by these scientists, the qualitative study of these 16 scientists is presented in this and the coming chapter. Broadly, the qualitative data was divided into three components: individual data, group data, and pedagogical understandings. This chapter presents the individual data that explores participant understanding. The data is presented by group and seniority in the following order: Senior Researchers at the Research Institute; Senior Researchers at the University; Junior Researchers at the Research Institute; and Junior Researchers at the University. For each group, two narratives are presented in detail. This data from the interviews provides a personalized perspective on the issues involved in learning to write and writing research articles in English. The next chapter presents the group data and the pedagogical understandings of these scientists.

6.2 Individual Data: Scientists' Narratives

Profiles of the scientists are presented in this chapter. The profiles are organized by the scientist's status within their respective institution, their publication history, and their comments regarding learning and participating in the writing of scientific articles. Scientists' narratives are presented in pairs with full details for each pairing.

6.2.1 Senior Researchers at the Research Institute

In this section, we present four senior researchers from the research institute. First, Dr. Alamo and Dr. Juarez will be introduced, and then Dr. Sanchez and Dr. Ortiz will be introduced in greater depth. Dr. Alamo completed his doctorate in France 20 years ago and joined the research institute immediately afterward. He is the first author of seven articles in English, three articles in Spanish, and co-author of a book in Spanish. He also appears as second author in two papers co-authored with graduate students. Dr. Juarez completed his doctorate 12 years ago in Mexico at the same institution where he is now a researcher, but he joined the institution 10 years before that as a laboratory technician. Unlike many scientists, Dr. Juarez has published eight papers as sole author in addition to the 15 he has published collaboratively. He said he typically collaborates with Americans or works alone. All but three papers are in English.

6.2.1.1 Dr. Sanchez[3]—Senior Researcher, Research Institute. Dr. Sanchez completed her doctorate in France in 1992. Like many senior researchers now in their 50s and 60s, she joined the research institute as an associate professor immediately after completing her master's degree. She undertook her doctorate while working as a professor-scientist at the research institute. She has been lead author on 11 scientific articles and co-authored nine others with graduate students, so she has been the primary writer of 20 scientific articles. Of these, 17 are published in English. In her research projects, she has collaborated with American scientists based in the U.S. and others based in Mexico. She has also led investigations with Mexican collaborators.

When describing how she learned to write scientific articles in English, Dr. Sanchez reported that she began learning English when she was a child. She attended a bilingual elementary school. Her father knew English and liked the language, and although her mother did not, both her parents were concerned that she and her siblings learn English, French, and Italian. She had exposure to English from a young age.

Later, when she entered university, she read science textbooks in English. "The books I bought were in English because they were cheaper [than the Spanish ones] and they avoided the bad translations . . . The scientific terms were not necessarily correct." She reports that from the beginning of her university studies, she read a lot in English. At this time she also began learning how to write a scientific research article, although writing in Spanish. "Each class had a laboratory and in the laboratory for each activity we had to follow the format of a scientific publication; there was introduction, materials and methods, results and discussion." She said she had some hardworking female classmates, and they often worked together on these assignments.

When she came to the research institute, she began researching and writing for publication. She submitted her first published article in Spanish to a bilingual journal. The journal's editorial department translated it into English, so it appeared in both languages. Thereafter, she began writing in English, "But it wasn't very difficult because I brought everything that was in my subconscious." Her years of reading English meant that, for her, writing in English was not "something traumatic."

Dr. Sanchez described her process of writing as a very solitary activity. She begins by reading a lot. "Sometimes I have some results—we are finishing a research project and I still don't know what the essence of the problem is. Then I read and with this reading comes the idea of how to focus the article." She said she once read a book about how to write scientific papers.

> It said you have to begin with results and when you have the results, you build the discussion and then the introduction. And I said, "No, I am going to write a story." . . . I create a scientific publication as if it were a story. It's as if I have an experience I want to share with you and I am going to say it in such a way that it will interest you . . . I begin with the title because this frames what I am going to write. I begin with the title [bangs on the desk to emphasize each idea], and the authors [bang], the institution [bang], and the introduction [bang]!

She described the process of writing one particular article. In this case, she was the lead author, and the second author was an American scientist based in Mexico. He introduced Dr. Sanchez to a particular methodology, and they extended it outside of her traditional area of research. She said that writing this article had been very difficult because the topic was unfamiliar. As stated by Dr. Sanchez, "I didn't know if I was going to be able to write it. . . . I had to read a lot, study a lot, synthesize a lot and write many versions of the manuscript." She said it probably took six revisions, and each time, it got better. "I would show it to the other researcher—'Is this okay?' 'Do you like this version?' and he would say, 'No, it's missing this,' 'No, there's too much here,' 'Again.' He was like my teacher." Even though this article went through many revisions, Dr. Sanchez maintains that writing a research article is about telling a story from beginning to end.

In another case, Dr. Sanchez described working with a master's degree student. The student had gathered some samples, but she was not certain about the quality or the data. Dr. Sanchez thought it was important that the student know if it could be published.

> The first time the manuscript was sent to an international journal it was rejected because the sample design wasn't good; there weren't sufficient samples. . . . Okay, fine, we're going

to send it to another less important journal, one for Southern California. It was well received and they published it.

According to Dr. Sanchez, it was very important to show the student that she had something to contribute and that she should not give up so easily. Sometimes "you try and they don't publish it but then you have to at least put in the effort to look for another option" for publishing. In this case, as in many others with master's students, Dr. Sanchez appears as the second author of the article. She explained that "psychologically it is much more important for her [to appear first], than for me to have a publication in a low impact journal."

Dr. Sanchez spoke about her strengths and weaknesses as a writer. She said that she genuinely likes to write, and she considers this a strength. She considers writing to be "very creative." She keeps a diary in Spanish. She has a strong ability to synthesize, so she can read many articles and know how to use what she needs. "It is a lot like a photographic memory, where I see what I read." All this scientific reading is in English. However, not having English as her mother tongue, leads her to doubt the clarity and comprehensibility of her texts. She always gives her manuscripts to one or two people to read before she even submits it to a collaborating colleague. "There is always a part of me that worries whether it's good."

She reported that in her current practice, she does not write in Spanish if she intends to publish in English. She explained that "you have to think in Spanish to talk in Spanish and you have to think in English to speak in English." She has colleagues who write in Spanish and then translate into English. She finds this unsatisfactory. "You read [their] text in English and it isn't good. You realize it isn't good, I mean it isn't the English that is expected for a scientific publication."

Dr. Sanchez also talked about the importance of being able to participate in international conferences. She stated that the ability to communicate findings in a conference presentation is as important as writing. Some of her students, she said, can more or less write in English, but they cannot speak it. They miss out on a lot if they cannot participate.

In summary, Dr. Sanchez learned English and French as a child in addition to her mother tongue of Spanish. As an undergraduate student in science, she relied on English textbooks and wrote a great many laboratory reports in Spanish that conformed to a research article format. Like many older scientists in Mexico, she did not pursue her doctorate until after she had a post as a researcher. When she did study, she went to France. As a researcher now, Dr. Sanchez writes in English and sees her articles as telling a story—a scientific story. Her collaborations are national and international, and her publishing experience reflects her international perspective.

6.2.1.2 Dr. Ortiz—Senior Researcher, Research Institute. Dr. Ortiz joined the research institute 18 years ago, immediately after completing his doctorate. He has served as both head of his department and dean of his division. His publishing record includes 24 papers, all published in English, and he appears as first author of seven. The majority of his papers are published with graduate students, and Dr. Ortiz appears as the second author. His papers have garnered over 100 citations, none of which are self-citations. Additionally, Dr. Ortiz has published more than a dozen technical reports and articles of interest to the general public regarding his area of specialization in Spanish.

Dr. Ortiz began learning English as a child attending a private school where he had American, British, and English-speaking Mexican teachers. As a teenager, he was sent to England for three months to learn English. One of the teachers was a man who was "very crazy, very nice, and he told us that immediately when we wake up we had to think about everything in English: 'I'm going to wake up, I'm going to shower, and then I will have my breakfast, blah, blah, blah.'" When he and his family vacationed in the U.S., they would often ask him while he was still a teenager to do the talking, so "I have never been afraid of talking in English," he said.

As a professional scientist now, Dr. Ortiz said that he has a method for writing an article.

> First, I do a "warm up" before I begin to write, by reading several articles. . . . I copy the style, the writers' style, looking at the journals, looking at the journals that you are going to submit to, what kind of language it has.

This preparation helps him with his greatest difficulty as a writer, his fear of the "white page." He describes the phenomenon as a lack of discipline where he procrastinates, answers emails, and invents distractions, all of which create a vicious circle that leads to greater fear. For him, the ideal writing situation would be a retreat where he has no distractions, but this is not achievable in his research life.

Dr. Ortiz said that he now often takes the role of an editor rather than a producer of articles. "With students, I try to edit and organize their things.... It's like being a coach rather than the athlete." He continued by saying that it is his philosophy to assist students and place himself as second author, even though he knows that other researchers put themselves first. Regarding students, Dr. Ortiz said, "I respect their authorship." Much of his own publishing is done with students.

Concerning establishing writing assistance of some kind, Dr. Ortiz said that he would establish a center with "two levels of service." One level would be to offer a week-long course for faculty in a place away from the institute. He said it should be in a nice place "with a nice garden [and] a library of literature of how to write science." In addition, he recommends a course for students, even though they have not completed their research, to teach them how to order and organize the information and graphics. He said the course would not focus on the "problem of language," but "how to write scientific literature in a certain way." It would be very important to title the course appropriately, he said. It should not be "'How to Write,' but rather, 'How to Improve Your Writing.'" The second level of service would be to have staff that can help with reviewing a paper; that person should have some level of knowledge of the natural sciences. This person would not provide translations but would try to improve the English of the writer.

In summary, Dr. Ortiz learned English as a child and solidified that knowledge with an exchange program to England when he was a teenager. He has no fear of English. He did his graduate studies in Japan, where the verbal interactions were in Japanese, but he used English for reading scientific texts and writing his dissertation. As a professional scientist, he often participates in international collaborations where the language of interaction is English, and he works with students to help them publish their work in international journals. He still feels his English writing skills are inadequate and would like to have a retreat as a place to compose articles.

6.2.2 Senior Researchers at the University

In this section we present the extended profiles of Dr. Zorillo and Dr. Lorenzo. Two other senior researchers at the university were also interviewed with the following biographical details. Dr. Gomez completed all his education in Mexico and began teaching at the university after he completed his master's degree 25 years ago. While holding the position of professor, he pursued his doctorate that he completed in 1993. His substantial publication record includes collaborating on almost 70 articles in peer-refereed journals beginning when he was still a student in his master's program. He appears as first author in half of the articles, and two-thirds of the articles are published in English. Dr. Rodriguez has been teaching at the university for 23 years after completing his master's degree in Britain. He also worked for a Mexican government regulatory agency and served informally as a translator for English-language foreigners who communicated with the agency. Ten years ago, he completed his doctorate at an institution in Mexico. He is the first author or second author with a student of eight articles, all published in English, and he authored chapters in three books in Spanish.

6.2.2.1 Dr. Zorrillo—Senior Researcher, University. Dr. Zorrillo introduced himself as "almost part of the inventory" because he has been with the university so long, close to 30 years at the time of the interview. He has participated as a co-author on more than 45 articles. He is first author of seven of those articles, and he appears as second author in several papers he co-authored with graduate students and thus had a significant role in shaping the articles. His research projects have received funding internally by the university, and externally through the federal government and international collaborative organizations.

Dr. Zorrillo completed his undergraduate studies in Mexico where Spanish and English activities were distinct. Some of the readings for his courses or relevant to his thesis work were in English. When he wrote his thesis in Spanish, his supervisor "made a ton of corrections, but I didn't feel that I had any particular deficit." He appreciated the corrections to his Spanish text because that professor had done his doctorate in the U.S., and "I figured that this [educational background] implied that he wrote well." However, students were not required to do much writing at the undergraduate level. "Disgracefully, this was the only document that I had to write apart from one or two documents that another professor asked for."

While working on his bachelor degree, Dr. Zorrillo also completed the university's series of six courses for learning English as a foreign language. He felt quite confident in his English abilities and obtained a score of 500 on the Test of English as a Foreign Language (TOEFL). That was the required score for entry into the American university graduate program he sought. But when he arrived in the U.S. to study, that level of competence was deemed insufficient. The university sent him to a summer-long, intensive English course before permitting him to begin his graduate studies.

During graduate school, Dr. Zorrillo had several focused programs to improve his scientific writing. "I realized I wasn't writing well," he said. It seems, however, that he was not the only graduate student who didn't know how to write in the expected scientific manner. Two professors in his program mounted a seminar in how to write reviews and present scientific papers. The course culminated by submitting an abstract for a conference presentation. "This was a very good experience," he said. Another course was developed because a professor said he was having difficulty obtaining grants and he determined that it was because he didn't know how to write grant proposals well. That professor organized a course in grant writing so that he could learn how to write them better, and the students, including Dr. Zorrillo, were allowed to participate. Through these courses of scientific and grant writing and making presentations, Dr. Zorrillo learned to write scientific texts in English.

In this period, Dr. Zorrillo saw increasing evidence of the importance of writing in his scientific life. First, he saw a professor attributing his difficulty in obtaining grants to his writing ability. He said, "I saw a direct relation between my classmates who wrote well, they were the ones that did well, even though they didn't necessarily express themselves well verbally." He said, "I became a little obsessive about trying to write well, and I still haven't succeeded."

As a writer, Dr. Zorrillo says that one of his strengths is his ability to put his ideas in order, and he credits the courses he took in graduate school for this. He learned a method of thinking about his texts that he continues to rely on. This method is a "script," and he must complete the script in much the same way as one would produce a movie, he said. The script has three parts: Problem, Approach to Solving It, and Relevance. By thinking about his manuscript in this manner, he creates a document that is clear and cohesive.

Despite his best efforts, Dr. Zorrillo continues to receive negative comments on the writing in his manuscripts. "Literally, they say, 'not good English' or 'the English is deficient,' and this is a big limitation." He has experienced a British reviewer criticizing American spellings, and vice versa, and the reviewers treat these spelling differences very seriously. "The reviewers have very little tolerance, in part because, unfortunately, none of them speaks another language."

His principal problem in writing is not having the words he needs when he needs them. "It slows me down a lot." At other times, he knows the word but not how to spell it. In fact, Dr. Zorrillo has a dictionary by his side when he writes. This problem of lack of vocabulary recurs whenever he has to write a letter or an email in English:

> It would take me five minutes in Spanish, but in English it takes fifteen, thirty minutes, and this is a serious problem because it involves so much more time. . . . Principally, it is because I can't find the right words.

Dr. Zorrillo is able to compose his articles in Spanish or in English, often depending on who his collaborators are. When his collaborators are American, he writes directly in English. When his collaborators are Spanish speakers, he often writes with them in Spanish and afterwards translates the text. "I don't do a literal translation," he explained. "I have another way of saying things not because I know a specific rule in grammar, but by how it sounds. I just begin it again in English."

Dr. Zorrillo has identified characteristics of Spanish science writing that are different from English science writing. He explained:

> Basically, the style in science is reduced. You can't say, "This is my style, respect my style." The golden rule is that you have to say what you want to say in the shortest and most clear way possible. If I want to describe a container or a glass or whatever, I have to find the shortest and most efficient way—that is the best style. Often in Spanish, we have a tendency not to be so direct, so efficient, so sometimes one adds more words. I think it is a cultural reason that we do this, thinking that we are more friendly in this way. . . . But in reality, normally, when one says things with few words, it comes across more clearly. So the principal correction [in English texts] is eliminating words.

Collaboration continues to be central to the production of science articles for Dr. Zorrillo. He has collaborated with American and Mexican colleagues, and international research teams that involved Chilean and Japanese scientists. Depending on the expertise of the research team, each person may offer specialized knowledge that is integrated into an article. Other times, the science expertise is similar, and rich discussions ensue concerning what is relevant to include in the article and what is not. The English-language proficiency of the co-authors varies as well. Sometimes Dr. Zorrillo finds himself having the most experience with English writing, and other times he defers to other collaborators who are native English speakers or simply more proficient than he is. This variable combination of scientific and language expertise makes each writing experience different.

The ability to use English as a medium of communication in the world of science creates opportunities that are not available to those who do not speak the language. The international collaborations have been important in establishing his reputation. "They have helped me by giving me the opportunity to enter other niches," he said. The situation is different for those who lack his level of English ability. He explained:

> I have colleagues that don't manage English and they have perhaps the same number of publications that I have. But I have double the opportunities available to me because of someone I know in the United States, [and] because I can express myself in English and writing in English. I have done international consultations . . . been invited to international forums and [scientific] councils because they could talk with me in English. It might not be good English, but good enough. . . . These are things that never would have happened only in Mexico.

Considering the assistance that would be useful in the local context for scientists to increase the number of articles they produce, according to Dr. Zorrillo, there are two types. One type of assistance would simply be a translation service, and this would be especially useful for people who say, "I need to communicate in English because if I don't, my international production won't be as good, but I am not interested in learning it." That person wants to give their article to someone, have them translate it, and this service would be paid for by the Mexican government or a scientific organization. The second type of assistance would be for people who want to learn to write better each time. For these people, a permanent office could be created "where I could call and ask for a word that I need in English." The other service this office could offer is,

> I write something to the best of my ability and send it to someone and they say, "Come here for half an hour and I am going to tell you, look, this should have been like this and this and this."

The idea would be for the scientist to come to learn and not just have a text corrected. "We try our best and it is so tiring—we all suffer with the pressure to publish." Dr. Zorrillo says that he has had problems with his writing and still does not feel very capable or secure in writing well.

Dr. Zorrillo added that another option to improve scientific writing is to put more emphasis on writing. He has given mini-workshops to fellow professors about incorporating more writing into science courses at the undergraduate level. He also occasionally offers an optional course in scientific writing in Spanish for graduate students. "We don't know how to write in Spanish . . . so it is a double challenge later to write in English." Despite the leadership Dr. Zorrillo has taken in bringing awareness of good science writing to the university, he said he still does not feel very capable or secure in writing well.

In summary, Dr. Zorrillo learned about scientific writing during his graduate studies with courses at the American institution where he studied. The crucial importance of writing became clear to him, and he worked hard to improve his English and writing skills that he lacked at the completion of his undergraduate program. He has a distinct "script" he follows to create the logic and coherence of his written work, but he still feels his English vocabulary is limited, and this is a source of frustration that prevents him from writing fluently. As a researcher, he has forged international and national collaborations, and he credits his English ability as one of the reasons for these opportunities. He continues to place great emphasis on writing and would take advantage of extra support if it were available because he feels he still does not write well.

6.2.2.2 Dr. Lorenzo—Senior Researcher, University. Dr. Lorenzo joined the university 20 years ago. As a researcher, he is a co-author of seven articles in peer-reviewed journals. He is the first author of four, and they are published in English. Additionally, he is the lead author of a book chapter and three articles published as conference proceedings. He reported that he has numerous obligations in addition to conducting research. "I have classes, coordination of distance education and I manage many students: I have a service project where they give classes to students of elementary school, classes about sciences. I spend a lot of time on this." In the last three years, he has supervised three undergraduate theses projects. "The effective hours of [research] work are outside of this," he said.

Dr. Lorenzo began his bachelor degree in Mexico and had to manage both English and Spanish: English for reading, Spanish for writing. He said, "Throughout the program from the first semester, everything we read was in English . . . and as we were advancing, there weren't books in Spanish." Where there were books in Spanish, they were bad translations. However, as he read in English, he would translate into Spanish in his mind. Thus, at this early stage in his scientific trajectory, he read in English but needed to translate it to comprehend and write in Spanish. Each semester there were 20 or 30 laboratory reports to hand in. "The professors are correcting everything—spelling, writing—and the writing in Spanish is a little freer than in English, which is briefer. One could make a report and make it like a story." This style is not permissible in English, he says. He continued writing scientifically in Spanish through his theses at the undergraduate and master's levels.

He went to England to do his doctorate, and that was the first time he was confronted with doing a major piece of writing in English. He recounts:

> I had already written a paper in English but with a lot of help from a colleague from the United States. So, ah, I had done a technical report in English, but they are not big things, they are little things; many letters in English, but a complete article in form and structure, no, no, I hadn't done it alone until I arrived in England.

In Dr. Lorenzo's case, his dissertation director did not provide much direction in science writing in English. "There wasn't much communication, not much direct communication." His dissertation director "just wanted the results." While in England, Dr. Lorenzo took a "writing English" course. He said, "I learned how an editor or a writer makes things from the different chapters" of a thesis into an article. He explained that the editor,

> expects to see something in the Introduction, and expects to see something in the Discussion, expects to see something

in the Conclusion: they expect to see something and if one doesn't put it there, it's bad . . . it's bad structurally.

Dr. Lorenzo has approached writing articles in different ways. In one instance, he transformed a student's thesis into a research article for publication. The student wrote the thesis in Spanish, and Dr. Lorenzo wanted to get it into print quickly. To do so, he made a line-by-line, direct translation from Spanish to English. It was structurally very bad, he said. After writing it, correcting it, letting it rest for some time, and then returning to it later, he was still displeased. "I still didn't like it because it seemed like Spanish writing." In retrospect, he believes it would have been easier to entirely rewrite the text rather than perform a translation. This would have also avoided the structural problems. Ultimately, a "style editor" from the journal that accepted the paper revised the manuscript. "Those changes were more grammatical, and some things that they didn't understand, he changed them." Dr. Lorenzo concluded that this process of writing "wasn't a good experiment."

In contrast to the experience of working with a thesis written in Spanish, the process that has been much more satisfying is collaborating with several researchers. "In science, one doesn't have to do things all alone," he said. "The people that have helped me to write are colleagues; people that I've written with together." Some of those collaborators are based in the U.S., and others are in Mexico.

In another case, Dr. Lorenzo wrote with a researcher in Mexico who is "very renowned, very good at writing in English and Spanish." Over a period of 15 days, for four hours a day, "we sat together, side by side, writing line by line. He would write a piece, we read it until we were in agreement, we continued again." They then gave the manuscript to another member of the research team "because we were saturated." This third researcher "made me see many things, many changes, changes of ideas, of structures, and this was his contribution." Finally, an American colleague who was part of the data collection process "was the moral support." That paper was accepted very quickly. Similarly, another paper—a collaboration with a noted American researcher—was quickly published, in part because "the whole thing was very focused since it was what [the American researcher] wanted."

When asked what would help Dr. Lorenzo in his efforts to publish, he answered, "Tutors, not translators!" He would want to have "people that understand well what a paper is" and someone "who could review it." He said it would be best to write "in direct interaction" with someone in "reviewing and correction."

In summary, Dr. Lorenzo reported that doing research and publishing is just one of his many obligations in his role as a senior scientist at the university. His most satisfying writing efforts have been collaborative with colleagues both in the U.S. and within Mexico where there was a great deal of interchange throughout the process. He still learns from his more experienced collaborators. Translating a text from Spanish into English was not a satisfying means of creating a manuscript although it was successfully published. His personal style involves more "story telling" and extensive descriptions that are more acceptable in Spanish scientific writing than in English, because English scientific writing demands he be direct and brief. The kind of assistance he would seek is a relationship in which he could interact directly with someone with more expertise in scientific writing in English.

6.2.3 Junior Researchers at the Research Institute

In this section, we present the extended profiles of Dr. Salinas and Dr. Carrillo. Two other junior researchers at the research institute were also interviewed with the following biographical details. Dr. Papel completed her doctorate five years ago in the U.S. She has published six English-language articles in peer-reviewed journals and written seven Spanish-language technical reports for Mexican agencies. Dr. Ilana completed her doctorate one year ago in Mexico. She has nine English-language papers in international journals and three book chapters. She is lead author on all but one.

6.2.3.1 Dr. Salinas—Junior Researcher, Institute. Dr. Salinas joined the research institute after completing a postdoctoral fellowship in Canada, and his doctorate in England. At the time of our interview, he had published four articles as the first author, one as second author with his doctoral student, and participated as a co-author in nine more papers dating from the time he was a bachelor's degree student. All his articles are published in English.

In Dr. Salinas's research writing development, his early introduction to "publishable science" was crucial. This introduction occurred when he was studying his bachelor's degree in Mexico. He began working in a laboratory where research was being conducted. He said,

"I began to do a little research and then, the results I was getting were results that could be published." All his interactions were in Spanish at that time, but he began to "gain experience in how to organize the material for writing an article, how to select it, and how to go about structuring it to write the article." There were discussions among the laboratory research team members about creating articles based on the work conducted there.

He credits his two supervisors, who already had publishing experience, with mentoring him. These mentoring relationships continued through his master's program in Mexico at that same institution. He said they provided a great deal of help, but

> the initiative was mine to begin to write articles. I told them that I have this, and I would like this in the article. I would like to include this part. Then if I didn't know how to put it, they helped me.

When Dr. Salinas was in his early 20s and working on his master's degree, he began intensely studying English. He said that all Mexican youngsters have English classes in high school, but it is rare that one really learns the language in that context. He had been reading science articles in English from his bachelor's degree days, but he did not have enough knowledge to write. He also decided he wanted to do his doctorate abroad. Therefore, he "dedicated one whole year" to studying English. This required two hours a day, five days a week, in addition to his activities in the university and the laboratory. For the young Dr. Salinas, this effort was essential so that he would have sufficient knowledge of English to be accepted in England for his doctorate.

While studying his doctorate in London, Dr. Salinas described deeply collaborative writing practices with his fellow doctoral students, postdoctoral fellows, and his doctoral supervisor. He recalls working on articles with members of his research group, who together published five articles in three years. He says:

> I remember it with great pleasure because we had the two postdocs: one who was very good with his hands and with experiments and the other that was good at doing calculations, making theory, etc. And the doctoral students, such as myself and my other classmates, when we wanted to discuss the research, the ideas in the laboratory, we did it with those two to

begin with, you know. So, this whole article would reflect the teamwork that was done in the group.

Dr. Salinas also described a process of writing one of the articles that came from his doctoral thesis that was collaborative. "I was still in the process of learning about how I could construct the final product, and I still needed the help of the people that had more experience than I." He became more independent and could construct articles himself. At that time, his supervisor "kind of played the role of referee" by vetting the scientific content and also "a little as editor" for commenting on the form and editing the manuscript.

Although now Dr. Salinas said he usually writes his articles directly in English, that was not always the case. He used to "think about the whole research project in Spanish. And if I wanted to write in English . . . first I would think in Spanish and mentally transform the ideas" into English. Mastering the language has been a gradual process. At first,

> I passed everything to my supervisor, my English classmates . . . In this way, I was learning to write articles and above all, English. . . . I had lots of people around me [and] for whatever doubts I had, I could go and consult with them.

After completing his doctorate and postdoc in English-speaking countries, Dr .Salinas finds it natural to think and write in English. Nonetheless, he still has doubts about grammar and logical flow of the "whole story." He described his experience of writing an article:

> Okay, say I have the experiments, I have my results, I have it very clear what it is that I want to say. Then I write it. Then, when I describe the technical part, the scientific part, in English, I know how I want to say it and how to put it. Grammatically, I can do it relatively well. The idea follows, you know? I want to put it together with what came before it. From the point of language, there it starts to be very hard. Then I need someone with more experience, you know, a native speaker of English, for example, to look at this part, who reads the final document and tells me, ah, look, move this from here to there and you can improve it.

When he completes a manuscript, and before he submits to a journal, he always has someone read it "to assure that the story flows well." He does not feel confident that what he writes will sound right. "Almost all the people that are in the department write articles in English and it's not unusual that someone goes and asks questions of them." On the other hand, he finds that his learning English as a second language also makes his English more accurate. He has collaborated with native English speakers who make errors that he can easily spot because of his formal grammatical training. For example, he cited one person who always writes "then" in place of "than," and Dr. Salinas corrected him.

Dr. Salinas's publishing experience includes one manuscript that was rejected and later accepted. He submitted it to a journal, and the referee criticized the scientific content and the English. But, he and his collaborator "didn't agree with what the referee was saying, and we sent it to another journal. There, the reviewer said absolutely nothing about the English." The journal that rejected the article was not any higher in prestige or importance than the one that accepted it. It is the only experience he has had of being told that the writing was bad. At this point in his career, Dr. Salinas only receives minor notations from referees about a manuscript's writing, and referees comment only on the scientific content. "I can publish now without major problems; if they don't accept an article from me, it is not because of the English."

In summary, Dr. Salinas experienced seeing real scientific research and writing while still an undergraduate student; this shaped his experiences as a scientist. He participated in deeply collaborative writing activities, and these practices have continued in his professional life. Although he did not learn English until entering university, his graduate and postdoctoral studies in English-speaking countries seems to have solidified his confidence and ability to write in English. Nonetheless, he continues to have doubts about the "flow" of his written texts. He often seeks informal affirmation and correction of his manuscripts before he submits them to a journal.

6.2.3.2 Dr. Carrillo—Junior Researcher, Institute. Dr. Carrillo completed her doctorate in Spain four years prior to our interview. She is now a researcher in the same institution in Mexico, where she obtained her master's degree. Her publication record to date includes ten published articles, and she appears as the first author of five. All the articles are published in English. In addition, she is lead author of 12 conference proceedings, and the majority of them are published in Spanish.

Dr. Carrillo began learning English when she was a university student, and not all her attempts were successful. She took English classes in addition to her science studies, but they were general English classes "for communicating with people in the street," and "I have not taken courses in technical English for writing articles." She said that when she was in Spain studying her doctorate, a course was available to students, but "my level of English wasn't very high. It wasn't high enough to allow me into a technical writing class, so I couldn't take it." She continued, saying, "It seems to me that [her current institution] gives some courses, but the majority are directed to students and, further, it is at a time when I cannot attend because I am always so busy."

The greatest influence on her writing has been reading scientific texts in English. Now, understanding English articles is not a problem because she was immersed in reading English throughout her academic career. When Spanish translations are available, "they aren't very good overall." Learning to read scientific English was also hard work and occurred without any formal guidance. Dr. Carrillo lamented that the science materials were substantially different than the language she was learning in the general English classes available to her. "Little by little, I got accustomed to reading articles, and since they are technical topics, little by little I got so that I could understand everything."

Dr. Carrillo described her attempts to transition from writing in Spanish to writing in English. "I have realized that the idea is not to translate. . . . I try to write what I want to say thinking about it in English. Before, I made literal translations in my mind." Sometimes she cannot find the words or phrase she needs to say something. When that happens, "I look at an article where I have seen a similar idea, then I look at the way they wrote it and take that structure." She relies on this strategy

> when I don't remember the order of the words or which word I can use in place of the one I am using because I repeat myself a lot; it's not very elegant or because it's not what I'm wanting to say . . . My vocabulary isn't rich enough.

This strategy of searching for words in others' articles is not always successful because when she reads that other article, she finds that "no, that is not what I want to say, it's something else."

The grammar and syntax of English still present difficulties for Dr. Carrillo. She mentioned the proper placement of adverbs, using passive voice (which is more common in English articles than in Spanish), keeping track of subject-verb agreement, and knowing which modal to use is still difficult for her.

Dr. Carrillo usually writes with collaborators. "I write the first version and the other person helps me to correct it, and then there's another person that helps us correct what we wrote." When she has a complete draft, she gives it to another co-author who "tells me, 'This doesn't go here, this yes,' and we discuss it: 'This should be included,' 'This should be filled out more.'" The third author makes the final review, checks the style, and improves the English. This collaborative writing and revision process has served her well in articles that she has co-authored with the directors of her dissertation. When they are in the same city, the co-authors sit and talk about the developing manuscript, but that is not possible when they are separated. Their feedback has been essential in Dr. Carrillo structuring and creating good scientific articles. Now, Dr. Carrillo is beginning to establish research and writing collaborations with colleagues at the research institute where she works.

During the process of multiple drafts and revisions, Dr. Carrillo says that the paper improves. Sometimes her co-authors change the text, and she is pleased because "that [new wording] is what I was trying to say. I agree with them at that moment." However, she said she looked back at some of those articles for this interview and found that "there are some things that I don't like . . . I try each time to do better."

Her papers in English often begin as conference presentations in Mexico. She presents her study and then writes it up for conference proceedings in Spanish. "Then these results I put into English trying not to do a literal translation." Developing an article generally begins in Spanish "until I get it to the point that my objective . . . is clear . . . we have to have the idea clear to be able to write in English." One difference between her first drafts and the final version is in knowing what is really necessary to include and what is not. "We end up making it shorter."

Dr. Carrillo is very aware of the ranking of different journals in her field. There are some journals that publish in Spanish, and the English journals have a range of rankings. Conferences that are held in the international, Spanish-speaking world are of a very high caliber. However, to share her results with the rest of the world, she must publish in English. Her strength, she said, is her ability to present mathematical matters and graphics in her papers. But, using English in her texts continues to be difficult. She said, "When we, people who speak Spanish, have to write, there are clearly some barriers, and it is very hard work."

In response to the question of what kind of assistance she would like if there were funds available, Dr. Carrillo named a number of activities. She would like courses in technical writing, although she does not have much time for such courses. She would also like to have a review of a completed manuscript and have it checked for format, style, and all the issues of grammar. She would like there to be someone to say, "'Okay, here what you are saying is not very clear.'" She would like to get together with that reviewer and say, "What I want to say is this, how is it written?" She does not want a translation of her paper, but to write it well in the manner she wishes to express her ideas.

In summary, Dr. Carrillo is immersed in a scientific field where not only all the literature is in English, but there are virtually no opportunities to publish in her own language. She came to English as a university student and has been working to improve her scientific English writing skills. As a junior researcher who always studied in a Spanish-speaking country, her collaborations are still mainly with Spanish-speaking collaborators. She is an active writer of articles and feels great pressure to establish herself in the scientific world so she can present her work in English at conferences and not only publish. English continues to be a significant barrier for Dr. Carrillo.

6.2.4 Junior Researchers at the University

In this section, we present the extended profiles of Dr. Ramos and Dr. Jornado. Two other junior researchers at the university were also interviewed with the following biographical details. Dr. Villa is a junior researcher who has participated in and contributed to published articles, but has not yet been the lead author. He primarily works in a laboratory and provides assistance to more senior researchers. As a junior researcher, Dr. Gonzalez has published three articles as lead author and is named as a co-author on seven more. All are published in English. He completed all his education including a postdoctoral posting in Mexico, and additionally he has taken a few courses in the U.S.

6.2.4.1 Dr. Ramos, Junior Researcher, University. Dr. Ramos studied his bachelor's degree at the same university where he is now, and he began teaching more than 20 years ago. Later, he completed his master's and doctorate at local institutions of higher learning. Dr. Ramos is considered a junior researcher because he completed his doctorate barely five years before this interview although he had many years of teaching and participating in research projects. He is the first author of five published articles, and another was under review; of those, all but one are published in English.

Dr. Ramos began studying English when he arrived at the university to study his bachelor's degree. He took general English courses offered by the non-credit Language Center, and has taken a few additional courses since, during periods when he did not have to teach. Many years later, he also took an eight week preparation course prior to taking the TOEFL exam. He pointed out that these were all general English classes and not writing classes nor specialized scientific writing courses.

Writing in general is difficult for Dr. Ramos. His master's thesis was only sixty pages long and his doctoral dissertation was less than 100 pages; both were written in Spanish. To accomplish these writing tasks, he was directed by his advisors. Despite the short length of those documents, he was able to publish four articles out of those research projects in collaboration with his advisors. He has learned to "write over time," and collaborating with others has been especially helpful. Nonetheless, he says, "I know that I don't do it well but I have learned. I wasn't taught how to do this kind of writing. It is hard work." He remarked that

> It is not easy to create an article: it takes a long time, sometimes years can pass and I haven't published and it has nothing to do with the language. You have to dedicate a lot of time, a lot of effort, many hours.

Writing in English is even more difficult. "Sometimes I try to write in English. . . . The thing is I have read a lot in English, a lot of scientific articles in English, so this should form some things in your mind, right?" He continued by saying, "All my ideas are in Spanish." For example, in producing one of his articles, he said, "There were a lot of things that I couldn't say in English, so then I wrote Spanish in some parts, and [collaborators] helped me to translate them." In another article he co-authored with two Mexican colleagues who had studied in the U.S., he said the reviewers had been very critical of the English, and "it was very embarrassing for me." The three authors then hired an English reader, and they shared the cost of that service. This cumbersome process also delayed publication of their article for more than a year. In another instance, his collaborator was a Mexican scientist who had spent four years in England doing his doctorate. They sat together making corrections to the language of their manuscript. Now he always budgets for translation services in his research projects.

Managing translations of his work has not been straightforward. He reported that he has used computer translations, but this strategy was not at all effective because the English required a great deal of correction. Several times he has given his manuscript "to someone who can read it and translate it for me. But, in this case, I have to re-read it afterwards because they see with eyes of a translator, not the eyes of a researcher." This dichotomy has extreme consequences, Dr. Ramos said, because "sometimes they change things, and I have to really focus on that."

Dr. Ramos talked about his strengths and weaknesses as a writer. He said that his strength is "wrapped up in research in such a way that the ideas are clear to be able to get the sense of the project." Emphatically, he said that he has many weaknesses. In particular, it is "so difficult" to begin to write in English. Yet, he continues to try to use English as the language of composition. He said,

> There are people who say, "Write it in Spanish, what are you writing it in English for?" But in my mind, I begin to write in English, but then, I go back and I try to write it in Spanish. It's a combination, and I dedicate many hours to putting it in English.

Another complicating factor for Dr. Ramos is the range of professional responsibilities he has. He said that in addition to any research he conducts and writes up, he has to give classes at the undergraduate and graduate levels, accompany students on field work, supervise students in their professional practice activities, and perform student academic advising services.

In summary, Dr. Ramos has focused more of his professional life on teaching rather than research and publishing. This situation is typical of older professors in the university who joined the faculty without graduate or doctoral degrees. The increased pressure and desire to publish is quite recent and coincides with the completion of his doctorate.

As an academic, he did all of his studies at Mexican institutions that placed little demand on writing in English. He relies on Spanish to compose his papers well and deals with translating them into English at a later point before submitting them. He continues to have many institutional responsibilities in addition to conducting research.

6.2.4.2 Dr. Jornado—Junior Researcher, University. Although Dr. Jornado joined the university as a professor more than a decade prior to our interview, he only completed his doctorate two years ago. He has published four papers as part of a research team of two Mexican colleagues and a Russian researcher based in Mexico. Dr. Jornado is the lead author of three of those papers, one of which is published in Spanish.

Dr. Jornado said that like many Mexican youngsters, English did not become important to him until later in life. English was one of his subjects when he was in middle and high school, "but I didn't learn anything." When he came to university to study science, he found he had problems because he could not read the texts in English. During graduate school, he decided to study English, but he only completed a six-month course. After completing his master's degree and joining the university's science department, he took a three semester course to improve his English skills. These courses "helped me a little to have a basic level of communication," he said. He continues to study English at the university language school. He recently took the International English Language Testing System (IELTS) exam, and scored in the "modest user" level.

In his graduate studies in Mexico, Dr. Jornado found that English was important for reading, but not for writing. All the readings and references are in English, but "we had no requirement to write in English." As a result of the English courses he took, he was able to understand the scientific articles he read. Currently, as a professor, he sees it as a service to his students to require them to read and understand textbooks in English.

Dr. Jornado's experience of writing science differs depending on the language he uses. He has written papers in Spanish for ten years, often contributing to conference proceedings and serving as a referee of proceedings. He also enjoys writing in Spanish for the general public and for children. But, English is a different matter. In English, "I have never written a paper alone." He writes in collaboration with colleagues. He continued, saying, "In practical terms, what I have done is checked the publications of other colleagues in English and I have checked how a graphic is discussed."

His research is principally conducted as a part of a team of four scientists. He explained:

> We do the study, an experimental study in the laboratory. We design experiments, take results, make graphics . . . and we discuss the results: First, I go back to my notebook—in general it is in Spanish—and I begin describing my experimental results. . . . But the writing is something we leave to the end.

For Dr. Jornado, the process of developing a paper is rich with discussions of the content of the experimental and theoretical work. Where they put the emphasis "depends on the journal we want to send [the paper] to."

The writing itself is often done among the colleagues in Spanish, and then they translate the text into English. None of the members of the research team has English as their first language, and one is a native Russian speaker. On occasion, they have used a computer program for translation. More frequently, however, they pass their manuscript among colleagues for comment and suggestion. As a result of his recent English courses, Dr. Jornado said that he tries now to write directly in English without first writing in Spanish. He reads articles and notices their style, especially those written by people who do a lot of publishing. He sometimes selects phrases to use in his own papers, especially, he noted, phrases to begin a paragraph. There is a list of thirteen phrases he keeps handy. When he does write in English, he often asks his wife to review it, explaining that she is an editor of a Spanish journal and has a lot of experience in writing. Further, her English is better than his, "so she helps me a lot and I have a lot of trust in her and her writing."

English proficiency continues to be a major issue for Dr. Jornado. Attending conferences conducted in English is difficult because "at best, I don't understand half" of what is said. The other half, "I imagine or I reconstruct it, and this has an impact . . . maybe I don't understand the main idea." On the other hand, "maybe because of my personality," he said he does not have problems communicating. "Even though I speak badly, I try, right, I try." He added, "One of my goals is to continue studying English."

In response to the question of how to assist him in his science writing, Dr. Jornado suggested improving the level of writing overall. He said that good writing is a problem in Spanish as well as in English. He suggested that there be contests, for example, of good writing that not only consider technical writing such as articles but writing as art as well. He also suggested creating an Internet magazine or journal that would provide activities to teach people how to write better. Workshops are useful, but he has had no time to attend any because he is so saturated with other work. He mentioned that a visiting professor had come to the university on a sabbatical, and she served as an editor for people's papers. That was very useful, but he said it would be better to have a permanent, formal service of support for professors' writing. This kind of guidance is what he offers to his students when they write their laboratory reports, but there is no formal system of assistance for the professors.

As a member of the science faculty of the university, Dr. Jornado said there is increasing pressure to publish research work, but this is in context of other obligations. He has courses to teach, administrative activities as the coordinator of the graduate program, and working with students in their own academic trajectory. These activities leave about 15% of his time for research. Doing research work has many important implications: "It supports my school, supports my career, and supports my students." When he demonstrates publishing productivity, it also has economic benefit to him personally through the university merit pay program and payments from the federal National Organization of Researchers.

In sum, Dr. Jornado continues to look for ways to increase his scientific productivity since recently completing his doctorate in Mexico. This productivity would garner him economic benefit, but it would also support the scientific efforts of his institution. However, he reports being stymied by the many institutional responsibilities he carries, and also by his lack of English. Although reading scientific English presents no problem, he has not yet composed an article in English alone. He is a member of a research team in Mexico that is productive, but language continues to be an obstacle in their publishing efforts. Further, he finds that attending conferences conducted in English is difficult because he does not understand "half" of what is said. Nonetheless, he presents his work in English and takes the risk of communicating as best as he is able with other conference goers.

6.3 Summary of Findings

Several trends stand out from these self-reports of scientists about their experiences in learning to write scientific articles and their continued strategies for doing so. First, for most participants, learning English was usually done in an extra-curricular manner, and their mastery never feels wholly complete. While pursuing their disciplinary studies, they additionally dedicated time to learn English. This was often through non-credit courses at the university where they attended undergraduate or graduate studies. Even the most prolific and experienced scientists continue to feel a degree of uncertainty about their writing, and this comes from the fact they are not writing in their mother tongue. Second, the scientists are cognizant of written scientific English as being different than scientific Spanish, although both share the same fundamental structure of introduction, method, results, and discussion. English is seen as more direct, concise, and concrete than Spanish. English writing loses the qualities of picturesqueness and flowery literariness that is valued in Spanish. Thus, writing a scientific article in English often requires thinking differently about what to write as well as how to write it. Third, relationships of collaboration permeate their educational and professional lives. Research is conducted in teams, and writing almost always involves more than a single author. As Dr. Lorenzo points out, "In science, one doesn't have to do things all alone." The collaborative process is often slow and involves many revisions. However, collaboration is almost always reported in a positive light, providing opportunities to explore the scientific content and the manuscript itself. Finally, the scientists make decisions about publishing their work in Spanish or English and the impact factor of the journals to which they submit their work. These decisions are based on the scientists' own sense of the importance of their research and their ability to position their work as being of interest to the international community. Several scientists lamented that publishing in Spanish or in Mexican journals would minimize the readership for their research and the recognition they receive as scientists.

The scientists' narratives presented in this chapter further support the statistical data presented in Chapter 5 that the four groups of scientists are somewhat distinct. In lived experiences and educational trajectories, the four groups of scientists (research institute, senior; research institute, junior; teaching university, senior; and teaching university, junior) have different emphases. The senior faculty at the research institute seem to have had an early grounding in English through home or elementary schooling. English seems more accessible to them and has contributed to their entry into the field of science through the facilitation of English reading and writing of scientific materials. The senior faculty in the teaching university seem to have had a much harder time in acquiring English for scientific purposes. The learning of English seems to have happened far later in their educational trajectory and in conjunction with graduate studies. Broadly, working with English seems far more of a struggle than for the senior faculty at the research institute. The junior faculty in the research institute are immersed in the need to publish in English but have not been properly prepared for this in their prior formal schooling in English. The development of English writing for the scientific purposes seems to result from graduate level interactions with advisors and other professionals. The junior faculty in the teaching university seem to be struggling with writing for publication in English. They write collaboratively with other faculty who have greater control over English writing and seem to be searching for practical solutions to help them publish research articles in English including the use of translation techniques and usage of translation services.

6.4 Final Comments

This chapter presents qualitative individual data that explicates the educational experiences of learning to write scientific articles in English as a second language. The approach taken in this chapter has been qualitative and individual, allowing access to the life stories and specific experiences of each scientist. In the next chapter, this same data set is explored as group data to allow a broader and more applicable set of trends to emerge.

7 Developing Scientific Writing Expertise: Qualitative Group Data

7.1 Introduction

In the last chapter, qualitative data relating to the interviews of individual scientists was presented. We presented detailed information from eight of the 16 interviews that were conducted. In this chapter, the whole data set of 16 interviews is used and referenced to develop collective group understandings. This chapter summarizes across the whole group the explicated educational events that contribute to the development of research writing. Some of this data is organized as a shared trajectory of learning events. This chronological organizational scheme allows the description of the common learning events which helped all these participants to acquire the ability to write research articles in English. To further understand this data set, this chapter also describes the trajectory of learning events for each of the four subgroups in this study: senior and junior scientists at the university, and senior and junior scientists at the research institute. The subgroup analysis is based on the findings presented in Chapter 5, in that there are differences in the degree of difficulty that each group feels in relation to writing research articles in English. In this section, the specific learning events that contributed to the English research writing development for each subgroup are identified.

In the next sections of this chapter, group pedagogical understandings are addressed directly. Participant scientists' statements from their interviews are considered in order to identify the educational interventions they would consider helpful or inhibitive in relation to their current scientific writing practices and development as writers. The last section addresses the affective factors involved in writing research articles in English as a second language. Taken together, the data in this and the previous chapters allow an informed discussion of the perceived needs of professional scientists in writing research articles in English as a second language. The ways in this can be facilitated to emerge is covered in the third section of this book.

7.2 Group Data: Trajectory of Learning Events

Having described the personal perspectives on issues of learning to write research articles in English as a second language in the last chapter, in this section, the qualitative data from the individual interviews is grouped and presented as shared information. For ease of reference to the individual interviews, the first time an individual scientist is referred to in this chapter, their status as a senior or junior researcher, based in either the university or research institute, is indicated in a footnote. The aim of this chapter is to organize the group data in a format that allows commonalities to emerge in relation to useful educational experiences. The analysis for this section consists of a careful re-reading of all the interviews and specification of learning events that contributed to the development of research article writing in English as a second language. These learning events are organized chronologically, addressing the formal progression of educational degrees from bachelor's degree to professional life as scientists. The outcome of this analysis, presented in Figure 7.1, consists of a shared educational trajectory that describes specific educational activities, according to educational degree, that contribute to the development of research article writing abilities in English as a second language.

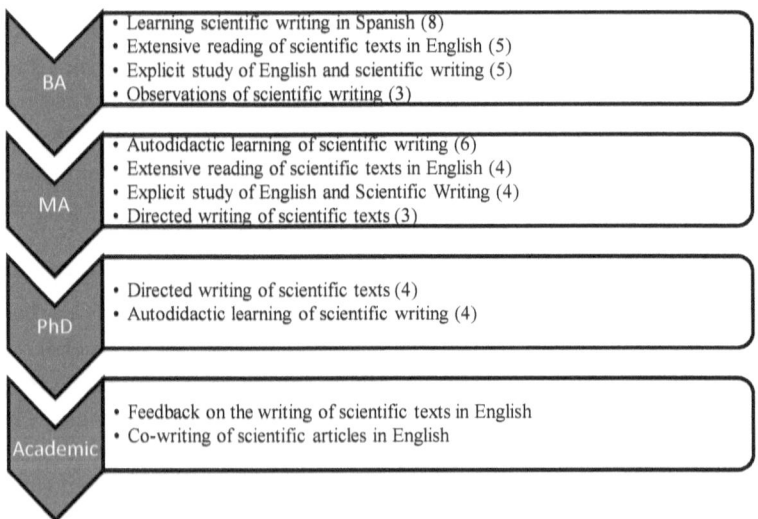

Figure 7.1 Progression of learning events through higher education and professional life (n=16).

7.2.1 Educational Trajectory: Bachelor degree

As can be seen in Figure 7.1, while they were undergraduates, the scientists in this study reported four distinct activities they believe developed their abilities to write research articles in English as a second language. It is important to note that all participants attended universities in Mexico, where classes were conducted in Spanish. As such, these specific facilitating learning events took place within that linguistic and cultural context. As specified by these participants, disciplinary science texts were read in English. There was limited access to translated science textbooks, but this was not particularly helpful, as specified by Dr. Sanchez,[4] who claimed that the translations were "bad." Furthermore, as specified by Drs. Rodriguez[5] and Lorenzo,[5] these translations were often not up to date, and were incorrect in places. For some situations, as reported by Dr. Juarez,[6] there were no disciplinary materials at all in Spanish. Accordingly, early in their academic careers, all these scientists stated that they read disciplinary, scientific material in English.

In a situation of this type, one would expect that English-language training would be integral to science education at the bachelor's level. This was not the case. For example, Dr. Jornado[7] reported that the requirement to read scientific material in English was very difficult because he had had no English-language training. Of this group of 16, 10 scientists began to explicitly study English as undergraduates. Typically they enrolled in general English-as-a-foreign-language (EFL), non-credit courses in the university. These language classes were not focused specifically on science or scientific reading or writing, but rather developed general English abilities. For Dr. Carrillo,[8] who had no English exposure prior to entering the university, these general classes were insufficient in attempting to read the texts of her discipline. It was through determination that "little by little I got accustomed to reading articles [in English] . . . so that I could understand everything," she said. As reported by Drs. Gonzalez[9], Lorenzo, and Zorrillo[10], in addition to these more general EFL courses, explicit courses/workshops concerning scientific writing in English were offered and taken. As expressed by Drs. Jornado and Ramos,[11] this extra-curricular language study was motivated by their need to read in English as a central part of their disciplinary studies. An additional reason for studying English in these courses was stated by Drs. Zorrillo, Salinas[12], and Lorenzo, who felt that their educational development as scientists may involve graduate study abroad, requiring high levels of English proficiency. These reports of explicit language learning of English suggest that this is a necessary response to the requirement to read disciplinary scientific material in English, and that this was not fully incorporated into these participants' educational programs.

As reported by the participant-scientists, learning to write research articles in English was facilitated at the bachelor's level through the experience of learning to write scientific genres in Spanish as part of regular disciplinary activity. Specifically, two scientific genres were developed: scientific notebooks and scientific reports. While these two genres were used in educational contexts, they do reflect scientific inquiry and scientific reporting and are part of a set of professional genres used by scientists in their research (Hanauer, Hatfull, & Jacobs-Sera, 2009). In relation to the writing of these scientific genres in Spanish, some participants reported that course professors reviewed these texts very closely and corrected spelling, grammar, organization, and structure of the students' texts. Thus, there was some feedback on the linguistic and textual aspects of writing scientific genre in Spanish. As reported, proficiency in these genres resulted from frequent practice where some students, such as Dr. Lorenzo, wrote as many as 20 laboratory reports in a year. The reports, the discussion, and the feedback on the reports were all in Spanish.

A final educational activity at the bachelor's level that was reported to contribute to the later development of writing research articles in English consisted of exposure to scientific writing aimed toward publishing by senior faculty. Some participants in this study, such as Drs. Salinas and Carrillo, were assigned to laboratories where senior scientists were conducting publishable research. As such, they became aware of the importance and process of creating a scientific article (even though they were not directly involved in the writing of these papers). Some of these experiences were particularly pronounced and influential. As reported by Dr. Gomez's[13] case, an American professor "inculcated in us the goal of writing scientific articles." As a result of this experience, he joined the professional association of his discipline and attended conferences while still a young student.

In summary, it is interesting to note the multilingual aspect of this educational literacy activity. As reported, at the bachelor's level, disciplinary scientific reading was, for the most part, conducted in English (L2), while disciplinary scientific writing was in Spanish (L1). Both literacy activities were seen as helpful to the development of later L2 writing abilities. The participants' reports suggest the need for English-language courses specifically directed at developing English abilities in reading the required scientific disciplinary material. Furthermore, exposure to real science and scientific publications produces initial understandings of the importance and processes involved in writing a scientific article in English.

7.2.2 Educational Trajectory: Masters Level

As can be seen in Figure 7.1, as master's level students, the scientists in this study reported four distinct activities they believe developed their abilities to write research articles in English as a second language. Two of these activities—extensive reading of scientific texts in English and explicit study of English and scientific writing—can be seen as extensions of educational processes that began at the bachelor's level. The other two activities—directed writing of scientific texts and autodidactic learning of scientific writing—can be seen as qualitative shifts in the process of learning to write a research article in English as a second language.

In relation to the activity of extensively reading scientific material in English, the main difference at the graduate level was the focus on scientific articles rather than textbooks. As reported by participants, disciplinary material at the graduate level consisted of professional research articles of which the vast majority was exclusively published in English. Thus, both the amount and the type of reading changed, offering the linguistic challenge of coping with large amounts of professional scientific material in English. This exposure is important as it involves comprehending the type of writing used in published research articles.

As seen at the bachelor's level, in master's level graduate studies, it is also important to explicitly study English and scientific writing. Once again, the predominance of English in professional publications, and the importance of these materials for becoming a scientist, make mastery of English literacy very necessary. However, as discussed at the bachelor's level, this was not always facilitated. For example, Dr. Carrillo reported that her university offered courses that developed scientific writing, but her level of English was insufficient to grant her access to them. Thus, at the master's level her English education involved general English that was not directed specifically at scientific reading or writing. Some participants, such as Drs. Zorrillo and Gonzalez, reported that they had opportunities to take workshops and seminars specifically directed at writing scientific articles in English as a second language. In other cases, the movement towards studying in English-speaking countries forced a process of accelerated language study. Dr. Zorrillo reported that after he was accepted as a student at an American university, he had to improve his English. Facing a similar situation, Dr. Salinas stated that during his master's program, and in preparation to attend a doctoral program in England, he studied English seriously. Overall, as with the bachelor's level, the explicit study of English literacy at the master's level seems necessary. For the current set of participants, however, this was not fully covered by existing educational opportunities.

A qualitatively different writing experience that contributed to the development of second language research article writing consisted of writing a master's thesis under the direction of a supervisor. This directed writing was often an iterative process, in which students wrote sections of the thesis and received commentary from their supervisor and committee. Most of the participants said that writing the thesis was the most extensive writing they did during their academic education to this point. The central role of reading in English is clear: their bibliographic references were almost exclusively papers written in English, even when they continued to write scientifically in Spanish (González-Reyna, 2010). Thus, the actual process of writing a thesis consisted of an educational literacy experience in which the conventions, requirements and forms of scientific writing were developed through direct feedback from a senior and publishing scientist. Furthermore, for the few participants who wrote their master's thesis in English, this involved direct work on professional scientific writing in English as a second language.

An interesting development reported by participants consisted of autodidactic approaches to learning to write scientific articles at the master's level. Based on the participant-scientists' interviews, this direction seems to result from three different forces: the requirement from the professional field for publishing, the lack of explicit educational resources (such as a writing center), and the decision on the part of the thesis supervisor to pass control of writing to the student-researcher. As might be expected, the articles and presentations that were written developed from the work conducted as part of thesis research. Several of the participant scientists reported making conference presentations and writing short articles for conference proceedings. These first articles of their career were sometimes written in Spanish and were usually co-authored with their supervisor, but the responsibility for the writing was given to the student. This put the student in the situation of being required to write a research article or presentation and to learn genre requirements. They held such responsibility without really having a serious support system to facilitate this process. Dr. Salinas said, "The initiative was mine to begin to write articles." The situation was more extreme for the students who wanted to publish in English. Several reported strategies of "copying" or "imitating" articles to learn to write, as stated by Drs. Ortiz[14], Idania, and Carrillo. The scientists also reported that they were self-taught. For example, Dr. Gomez said, "I am a self learner; I learned on my own."

Overall, at the master's level, we see a continuation of the need for focused language programs directed at developing scientific reading and writing in English as a second language. These courses would offer support to the extensive, professional reading required and provide a starting point for the writing of a thesis or professional scientific publication. At this level, we also see the role of the thesis supervisor in developing in their students the types of literacy skills and knowledge required to become fully functioning members of the scientific community (Ynalvez & Shrum, 2009). Finally, there is a movement in which the responsibility and requirement for quality scientific writing is passed to master's level science students. Without other institutional support, this was managed through self-study and the direction of thesis supervisors.

7.2.3 Educational Trajectory: Doctorate Level

As might be expected, entry into doctoral studies, with its associated enhanced professionalism, involved another qualitative shift in literacy practices. In the case of doctoral level studies, literacy learning activities became more focused on the production of texts. Two types of scientific texts were specified as of significance at this level of education: the doctoral dissertation and research articles. In both cases, there was a degree of writing direction provided by dissertation supervisors and other members of the doctoral committee. However, as reported by these participants, much greater and closer support was provided for writing the dissertation than for article writing.

The participant-scientists in this study wrote dissertations that were closely directed by their supervisors. As a public and scientific document, oversight of the dissertation included consideration of the quality of the scientific content of the work and its written aspects, such as the organization and structure of the document. Within the current group of participant-scientists, nine completed their doctoral studies in Mexico or Spain; the other seven did their doctoral studies outside of Mexico, but not necessarily in English-speaking countries (such as Japan and France). Of the 16 scientists, half wrote their dissertations in English, seven in Spanish, and one in French. With the exception of Dr. Lorenzo, who reported that he had little interaction with his graduate supervisor, the interaction with their supervisors was crucial to their scientific trajectories. Thus, for the scientists in this study, writing the dissertation involved personalized feedback and instruction relating to the scientific writing abilities.

Often, the dissertation was written with the intention of later producing one or more publishable articles. While this is an extension of doctoral work, the degree of support provided for this writing task was different than writing a dissertation, even though the same dominant figures (such as the doctoral advisor) were involved. As reported by the participants, writing for publication was very much an interactive process in which their first independent attempts were commented on and corrected by their supervisors. For example, the scientists in this study specified that they had to learn what kind of information to assign to each section of an article as they attempted to compact the lengthy form of a dissertation into a few pages for a research paper. Dr. Papel[15] described writing drafts and her professor disentangling her text that "went in circles." In the terms used by Dr. Lorenzo, they had to adjust their "story telling" so that it was appropriate for a research article. Dr. Gonzalez said that all his first articles were written with the direction and support of his advisor. The resultant articles were always co-authored by the student's supervisor, and in some cases, other members of the supervising committee. The doctoral student tended to work alone and then received advice from the supervisor. Accordingly, there was supervisor assistance; but the onus and the lead writer on this task was the doctoral student. (See Li , 2006, for a case study of a similar situation for a Chinese physics student). In a few cases only, doctoral level writing was much more autodidactic. Dr. Rodriguez, for example, said that he collaborated with a fellow graduate student on his first article at a time when "neither of us knew how to write an article." Dr. Lorenzo received little direction. Accordingly, some scientists reported that the work of writing a research article in English was mainly conducted alone as an autodidactic process with little help from their supervisors.

In summary, at the doctoral level, we see a shift in focus towards dissertations and research article writing, but there is a continuation in the educational methods used to facilitate the writing of these texts. As reported by participants, writing a dissertation involved getting direct feedback on the linguistic and textual aspects of the written text. As such, the process of directed writing seen at the master's level continued into the dissertation. Responsibility for writing research articles was placed upon the doctoral student with help from the doctoral supervisor. Often, particularly the younger scientists reported an iterative process of drafting text and receiving feedback and corrections from their supervisors. This directed, interactive process was especially evident when the participants described how they transformed their dissertation research into English-language scientific articles. However, as reported by some participants, there was not enough support, and as such, they contended with this writing process themselves and tried—in autodidactic manner—to develop their research article writing abilities. This, once again, continues literacy learning processes that were first encountered at the master's level.

7.2.4 Educational Trajectory: Professional Scientists

Learning to write research articles in English as a second language does not end with the completion of a dissertation; rather, it continues into professional life. As might be expected, entering professional life as a scientist brings new ways of learning to write scientific articles. For our participants, two means of literacy learning were significant at the professional level: collaboration during the research article writing process and learning from article reviewers' comments after submission. Both of these learning opportunities are directly connected to the central task of professional scientists—the production of published research articles.

As reported by our participant-scientists, the process of collaboration in writing research articles had different forms and purposes. (See Lillis and Curry, 2010, for an account of the networks afforded by collaborations.) One form of collaboration involved a continuation of the relationship of dissertation supervisor and student but on a more egalitarian basis. Collaboration in authoring permitted the young scientists to write with their former doctoral supervisors as junior peers. For example, Dr. Salinas and Dr. Carrillo, who are both junior scientists, continued to co-author with their supervisors for their first several published papers. The importance of this type of learning activity was also mentioned by senior researchers (Dr. Sanchez, Dr. Ortiz, and Dr. Zorrillo), who discussed offering this co-authoring arrangement with their graduate students as a way to mentor those students and young researchers into the field.

Other important collaboration and co-authorship relationships were also mentioned. Scientists who have English as a second language may collaborate and co-author with native English-speaking scientists to further understand English research writing and make sure that the final paper is acceptable from the perspective of the language and style used. As reported by Dr. Lorenzo, this form of collaboration can provide insights into particular turns of phrase and the structure of the research article. Dr. Alamo[16] reported that he always asks for an informal review from colleagues, especially ones who speak English:

> Other researchers feel very confident of what they write and they don't ask for opinions . . . But I don't feel that sure with what I write and I prefer to polish the manuscript before sending it to the journal.

However, collaboration was not limited to English-speaking scientists. Some participants mentioned the value of co-authorship and collaboration with other scientists (Spanish and other language groups) as allowing fruitful discussions on scientific content and presentation to emerge. Drs. Jornado and Papel reported that this form of collaboration included explicit discussion of the linguistic and textual aspects of the shared, emergent article, including issues of genre and style.

Following the submission of a research article for publication, reviewers' comments were also seen as opportunities to learn more about scientific writing. The participants in this study pointed out that reviewers' comments often included suggestions and recommendations for changes that addressed the form of the article and not just its content. (See Fortanet, 2008, who also contends that reviewer comments are intended as requests for improvement.) These comments could address structure, language, and style. This type of commentary is not always seen in a positive light (Englander & López-Bonilla, 2011). For example, Dr. Zorrillo's expressed his belief that undue criticism is directed at inconsistencies of British versus American spellings. Other scientists reported that reviewer comments have been helpful in improving their own scientific writing abilities. As Dr. Gomez stated, "With this help that they give us, we continue learning the form that they want."

In summary, analysis of the interview data with 16 Spanish-speaking scientists demonstrates that they experienced a consistent trajectory in their learning to write scientific articles in English. As shown in Figure 7.1, and as explained above, there are four distinct stages in the scientists' academic development: bachelor's, master's, doctoral, and professional. Each stage offers distinct opportunities for learning. The personal experience of each individual varied, of course, but the trend of particular practices at particular stages is largely consistent among the 16 scientists interviewed. This consistency in trajectory has implications for how to best create learning events for scientists whose first language is not English and the points at which specific forms of intervention may be appropriate. These pedagogical conclusions are presented fully in Chapter 8 of this book.

7.3 Trajectory of Learning Events within the Subgroups

In Chapter 5, and based on the analysis of the survey responses, four subgroups were identified as statistically different; accordingly, in this section we further analyze the interview data regarding the trajectory of learning to write science articles in English. The four defined subgroups consisted of senior scientists at the university, junior scientists at the university, senior scientists at the research institute, and junior scientists at the research institute. In this analysis, the specific literacy learning events that were emphasized for the development of research article writing in English for each subgroup were differentiated. This further analysis reveals that there were different foci in the specific learning activities at different stages among the four subgroups, suggesting possible differences in the development of scientific writing. Figure 7.2 presents a summary of these subgroup differences and their progression.

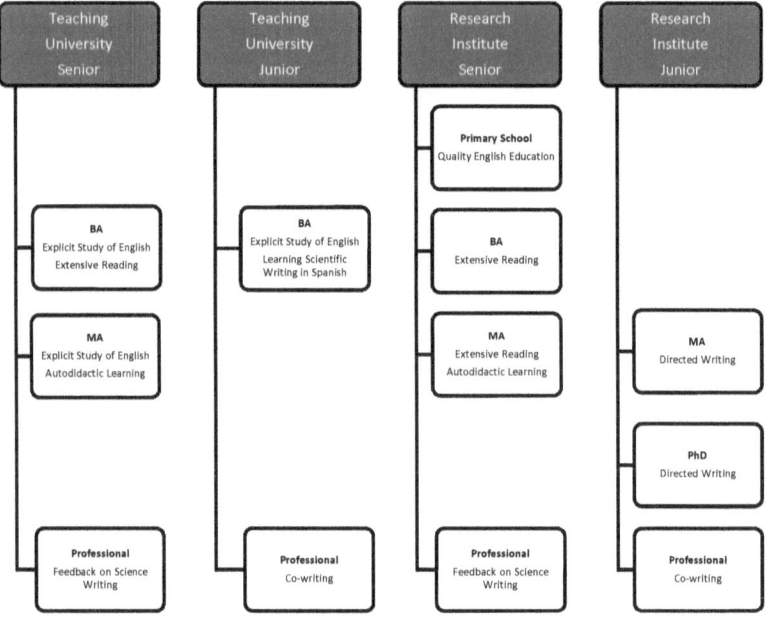

Figure 7.2 Progression of learning events through higher education and professional life by subgroups (n=16).

As can be seen in Figure 7.2, there are several differences of interest among the four groups of scientists. First, as reported in Chapter 5, the group with the least burden in writing research articles in English as a second language—senior faculty in the research institute—were also the only group to report on quality, early exposure to English as a factor contributing to their later development as writers of research articles. The senior scientists at the research institute started their trajectory toward professional writing in English at an earlier stage than all the others. Three of the four scientists in this subgroup, Drs. Sanchez, Ortiz, and Alamo, all were exposed to English as children or teenagers. In some cases, they studied at bilingual elementary schools, some went on international exchange programs, and others had family members who were fluent in English. These self-reports on the presence of early English exposure cannot be seen as a causal relationship that directly develops scientific writing. However, when one considers the reports on the importance of English reading and writing of scientific materials during all stages of higher education, and the relative lack of support for these second language literacy activities, it is reasonable to conclude that high levels of English at an early stage could facilitate and ease the process of learning science. It is also noteworthy that the only other scientist to mention learning the language at a young age—Dr. Papel—is also a scientist at the research institute, though he is junior faculty.

A second finding of interest is that both the junior and senior faculty at the university reported the importance of explicitly studying English in helping them to develop scientific writing abilities. This is possibly related to the previous finding in that the lack of early English education and the requirement to read scientific material in English in the university required some form of educational solution. Accordingly, it is not surprising that these scientists sought out explicit English instruction to help them develop within their chosen profession of science. Unfortunately, as reported in the last section, these courses were not always directed at reading and writing scientific materials and, in many cases, involved more generic English courses.

The first two findings suggest a different route for the development of scientific writing in English as a second language. The senior research institute faculty (and perhaps some of the junior research institute faculty as well) began higher education with enhanced levels of general English literacy that facilitated the process of both becoming a scientist and ultimately producing publishable research articles. University faculty, on the other hand, had to take English courses to acquire knowledge and ability in English in addition to learning science and scientific writing in English. Since not all the English courses were particularly suitable and beginning levels of English may have been low, this suggests a quite serious barrier to overcome in actually becoming a scientist. There is a double burden for these scientists: to learn the disciplinary content of their field, and learn that content in another language for which there is insufficient support. One clear ramification of this situation is the requirement for quality English literacy courses directed at scientific reading and writing for higher education students in the sciences.

A third finding relates to the differences between junior and senior faculty in later modes of learning scientific writing. The senior scientists at both institutions discussed learning from journal reviewer feedback, while the junior scientists at both institutions talked about learning through co-writing and collaboration. In both cases, the literacy learning process involved interacting with the wider scientific community and around the writing of scientific texts. There is a significant difference, however. For the junior scientists, particularly those at the research institute, the learning is more of a process of a writing apprenticeship in collaboration with more senior faculty, such as thesis or dissertation supervisors, and involves a process of directed writing. As junior professionals, their first publications were invariably co-written with members of the graduate research committee. This co-writing experience helped them continue to learn about science writing in English. The senior faculty actively collaborated and co-wrote with others both within Mexico and internationally. In addition, the senior faculty reported that reviewer feedback after article submission and review contributed to their learning to write scientific articles in English. For the senior scientists, ongoing literacy learning also resulted from peer interaction with scientists of similar levels of expertise, where they discussed what was appropriate, what was not to be included, and how an article could best be presented. Thus, more senior scientists continued to learn how to write a research article in English as a second language, but from their peers within the field and in response to a particular article.

Finally, Figure 7.2 indicates that there is some evidence of historical changes in science education in the different group trajectories for these participants. The older and more senior scientists at both the research institute and university reported autodidactic approaches to learning scientific writing. Junior faculty seemed to have more directed writing instruction from their thesis and doctoral supervisors and co-writing of research articles experiences with senior and researchers. There is still a need for autodidactic learning of science writing, but this is in coordination with more of an apprenticeship model of learning to write science. It is possible that master's and doctoral supervisors who are currently working may be taking a more direct approach to the writing of their graduate students. Thus, junior scientists currently feel that they did not have to teach themselves as much as did the more senior faculty. Senior scientists reported more often that they needed to learn how to write science on their own and less often that their first articles were co-authored with their dissertation supervisors. Thus, the mentoring relationship between professors and their graduate students may have strengthened over the past 25 years.

These third and fourth findings in relation to subgroup differences suggest the importance of individualized writing support for the development of scientific writing in English as a second language. Support may be in the form of directed writing instruction for the thesis or doctoral dissertation, co-writing with a more senior scientist, or later in professional life, co-writing and discussion with peers and responses to reviewers' comments. In all cases, this mode of learning to write is highly contextualized to producing specific, scientific documents within the setting of professional, scientific work.

Taking into consideration both the group and subgroup trajectories of educational development, an interesting model of required literacy learning intervention, emerges. At the beginning of higher education—at the bachelor's and master's levels—there seems to be a need for explicit educational courses directed at developing reading and writing skills for scientific materials. At the level of master's and doctoral studies, the model changes from group courses into individualized instruction in the form of an apprenticeship with a more senior scientist. Finally, at the professional level, scientific writing develops through peer discussion and response to reviewers' comments. Figure 7.3 summarizes this three-tiered development.

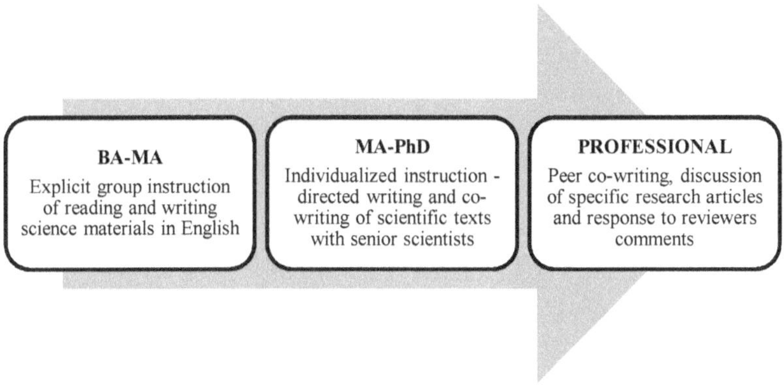

BA-MA	MA-PhD	PROFESSIONAL
Explicit group instruction of reading and writing science materials in English	Individualized instruction - directed writing and co-writing of scientific texts with senior scientists	Peer co-writing, discussion of specific research articles and response to reviewers comments

Figure 7.3 A schematic representation of the valuable educational experiences that develop the ability to write research articles in English as a second language.

7.4 Pedagogical Understandings: Participant-Scientists

As part of the interview process, each of the scientists in this qualitative study was asked about the type of pedagogical assistance that would be valuable in facilitating and easing the process of writing and learning to write research articles in English as a second language. The question was presented in relation to the possibility of a well-funded future option. The idea behind this question was to elicit the participant-scientists' understandings of pedagogical options best suited to their particular situation and educational history. Perhaps not surprisingly, their responses are very close to the analysis of the educational events they specified as significant and were reported in the last two sections of this chapter. Basically, participants support the usage of explicit language courses dedicated to learning scientific reading and writing and the development of some form of individualized instruction to help with the process of writing a research article. Of the 16 participants, 11 specified the value of providing English-language courses designed for scientists; 9 specified the importance of providing individualized instruction and support in relation to writing research articles.

It is important to note that two participants based in the research institute, Dr. Sanchez and Dr. Papel, felt that they had no need for any form of English writing support. This is consistent with the quantitative findings reported in Chapter 5 that 29% of the researchers at the institute said that language presents no barrier to their publication efforts. However, 14 of the 16 scientists in this qualitative study (87%) said they would like to have the option of some form of assistance in relation to the language issues involved in writing a research article in English as a second language.

In relation to the specification of the content and form of possible language courses that could be offered, a range of ideas were presented. Under the heading of explicit language courses, the scientists' suggestions included special three-day retreats, week-long courses, applied linguistic seminars, mini-workshops, full workshops, and a series of language courses including general English, technical writing, scientific writing, and writing for the popular press on science issues. Several scientists suggested setting up a "special department" that designed and offered courses such as these on a regular basis and in different formats. According to these scientists, there is a range of topics that need to be covered in such courses. In relation to scientific writing, suggestions included addressing order of information, textual organization, graphic presentation of data, and overall coherence. Recommendations for the content of these courses went beyond scientific writing in English. Some scientists specified the need for courses that develop scientific writing in Spanish. Others asked for English courses that developed verbal skills, such as the ability to present a scientific paper at a conference. Overall, there seems to be broad acceptance of the idea that explicit, group-based instruction of scientific writing would be valuable and that this needs to address first and second language scientific writing and also the textual levels of writing.

The scientists expressed the desire for individualized instruction in relation to a range of core concepts: writing centers, tutors, language reviewers, and a "personal trainer" for scientific writing. While the specific terms changed, there was a lot of agreement over the role and nature of this individualized instruction. Basically, the scientists in this study expressed the wish for on-going, easily accessible, individualized language support for their research article writing. As articulated by Dr. Carrillo, she would like to have a person who could review a whole manuscript, check it for format, style, and grammar, and tell her where it is "not very clear." She would like to "get together with that reviewer and say—'what I want to say is *this*. How is it written?'" In other words, Dr. Carrillo would like to work closely with an expert reviewer who is knowledgeable of her first language as well as English and could provide specific feedback on the specific article being written. Dr. Ramos expressed the same desire when he stated that he would like "the chance to sit with someone who can review what I write," and that this would "enrich the process" of writing and facilitate learning. Dr. Zorrillo suggested creating a special writing service office "where I could call and ask for a word I need in English," or to where he could write an article to the "best of his ability and send it to someone and they say, come here for half an hour and I am going to tell you, look, this should have been like this and this and this." As specified by Dr. Zorrillo, the idea is "for the scientist to come and learn and not just have a text corrected." While expressed in different ways, the overall suggestions for individualized instruction are reminiscent of a writing center approach. This involves the creation of a site with experts in scientific writing who have at least some knowledge of a scientist's first language (L1), and where there is easy access to personnel who can read scientific manuscripts and who have time to meet, discuss, and interact with scientists around their writing.

As specified by these scientists, the content of this individualized instruction relates to all aspects of the research article writing process. Dr. Rodriguez, for example, would like access to someone who could help with "corresponding and responding to editors and reviewers." This is in addition to more basic roles of "reading and reviewing a document." The suggestion is that this tutoring approach would address the wide range of academic research activities required of a scientist, including writing research articles, reviewing manuscripts, interacting with journal editors, understanding reviewer comments, and responding to reviewers. This raises an issue in relation to who is qualified to be such a tutor. Several of the scientists addressed this. Dr. Ortiz specified that the scientific writing tutor needs to have "some level of knowledge of the natural sciences." Dr. Jornado referenced a visiting professor who fulfilled the function of editing other scientists' papers as an example of a very useful interaction. Furthermore, some participants made it clear that they would prefer to discuss their research and article writing in Spanish (their L1) even if the text being discussed was in English. The basic understanding is that to provide the type of support needed for individualized writing instruction in relation to scientific tasks, the tutor would need to be a writing specialist with scientific knowledge and have the ability to converse with the scientist in their L1.

One scientist suggested creating a "translation service" for those who need to produce scientific articles in English but do not wish to learn English. This option is most appealing for senior researchers who need to publish in English but have little interest in attempting to learn the language well enough to write. This suggestion of translation was explicitly negated by three other scientists who stated they wanted to have control of their own texts and not simply have them translated. The overall direction was to provide individualized tutoring designed to help the scientist learn, understand, and improve their texts so that they say what they want them to say and in the manner expected by their contemporaries in the scientific community. As Dr. Ilana[17] expressed it, "so that things are said in the right way." Other suggestions included using corpus linguistics to specify language clusters of value for scientific writing, and to function as an aid to scientists writing English as a second language. It is important to note that the scientists in this study really felt the need for the creation of some form of writing center that offers them support. As stated by Dr. Jornado, students are offered guidance from faculty, but "there is no formal system of assistance for the professors." Or, as stated by Dr. Zorrillo, a "permanent office could be created" that offered writing services. Overall, the recommendations for educational intervention made by scientists in this study corresponded to the analysis of educational events that appeared in the last two sections. They have expressed the need to focus on writing and language skills by providing explicit language courses and for the creation of a writing center that provides individualized instruction from a knowledgeable tutor around their writing for scientific publication.

7.5 Affective Responses

The main aim of this chapter is to provide a basis for developing effective pedagogical approaches that can facilitate and ease the process of writing research articles in English as a second language. Up until this point, the chapter has addressed the types of educational events and pedagogical interventions that our groups of scientists feel are, or may be, beneficial. But, it was notable in analyzing the interviews that beyond the specific educational histories and pedagogical ideas expressed there was a definite emotional aspect to the description of learning and writing research articles in English. (See Belcher and Connor, 2001, for other multilingual scholars' first-person accounts of the affective element of writing for publication in English.) The qualitative data presented in Chapter 5 clearly shows that there is anxiety in scientific writing in English. While anxiety or other emotional responses are not directly part of a pedagogical program, there is no doubt that affect is an aspect of the experience of learning. Accordingly, in this section we address the characteristics of the affective responses noted in the interviews.

Specifically, two affective matters—anxiety and risk taking—were noticeable in the interviews with the scientists. Anxiety was identified in comments such as Dr. Salinas's who said he does not feel confident that what he writes will sound right; or comments from Dr. Gonzalez, who said, "I am still pretty unsure of myself." Dr. Ilana said, "I don't have confidence," and "I know what I want to say but I don't know how to say it in English." Other scientists spoke of writing in *Spang-*

lish—a combination of Spanish and English that would not be clear or acceptable as scientific English. Statements such as these reflect insecurity and anxiety in relation to the high stakes issue of research publication on the part of these scientists. To quantify these responses, the interviews were analyzed for comments expressing anxiety about writing in English. These data were organized by subgroups and are shown in Table 7.1. As can be seen in the table, both subgroups of junior scientists produced many more statements of anxiety than the senior scientists. At the university, the junior scientists expressed anxiety twice as often as the senior scientists. At the research institution, the junior scientists expressed anxiety three times as often. In total, there were more statements of anxiety on the part of the university scientists than the research institute scientists. These results correspond to the results found in the quantitative data presented in Chapter 5.

Table 7.1 Frequency of occurrence of scientists expressing anxiety or risk-taking concerning publishing scientific articles in English. N=16.

	Anxiety	Risk-taking
TU Senior	9	6
TU Junior	18	7
RI Senior	5	4
RI Junior	15	7

A different response to this core sense of anxiety were statements that expressed the need to take risks when writing in English. Risk-taking was expressed through comments such as, "I dared to do it: write in English" (Dr. Papel), or "My advisor helped me to not be afraid" (Dr. Ilana). As is seen in Table 7.1, occurrences of risk taking comments were not as frequent as those of anxiety. The senior scientists at the research institute felt that English writing involved the least anxiety and the least risk-taking among the four subgroups. The junior scientists feel the most anxiety and risk-taking.

In addressing this analysis of statements of anxiety and risk taking, it is worth considering the results reported in the last section dealing with the recommendation from these scientists that literacy courses and an easily accessible writing tutor would be helpful. Enhanced scientific writing in English, and the option of consulting with a scientific writing expert would both reduce the presence of anxiety and increase the willingness to take risks in writing scientific articles in English. These results, once again, suggest the need for some form of educational intervention and support for the scientific writing of these scientists.

7.6 Summary of Qualitative Results

The results of this qualitative inquiry into the educational trajectories and personal understandings of useful pedagogical interventions provide evidence of the needs and desires of this group of scientists in relation to the writing of scientific articles in English as a second language. The results on the individual and group levels suggest a very specific progression of educational interventions, and suggest that really interacting with the imposed requirement that scientific research be published in English means a long term commitment and the allocation of funds on the part of universities, research institutes, funding agencies, and governments. The importance of such multi-faceted and long-term commitments is also discussed by Salager-Meyer (2008). The results presented in this study make it clear that the necessary educational intervention is far from being a one-time course or some other quick fix. It consists of a concentrated effort over the full period of the training of a scientist into the continued support of professional scientists. As seen in the results of the current qualitative study, and presented schematically in Figure 7.3, we can think of this process as consisting of three distinct phases: (1) bachelor's and master's; (2) masters and doctorate; and (3) professional life as a scientist. During the early stages of higher education, there is the need for explicit courses that deal with scientific reading in English and scientific writing in both the first language and English as a second language. These courses should develop the necessary literacy skills to really understand scientific material written in English as a second language and begin the process of writing scientific texts in their first language as well as in English. First language science writing may be particularly important, as it can develop core understandings about the textual, procedural, and functional aspects of science writing, many of which may be transferable to English science writing.

During the master's and doctoral stages of higher education, there is a need for more individualized support of the process of writing scientific texts. Specifically, this is in relation to the writing of the formalized professional texts, such as a thesis or dissertation, and later in relation to research articles. Support is provided by thesis and dissertation supervisors, and it is possible that these supervisors need to be trained in how to focus on and enhance their students' writing abilities. In any case, some form of writing center support needs to be provided for students so that they can get writing support from a literacy specialist. This is in addition to the provision of group courses designed to teach the aspects of scientific writing at the graduate level. Thus, during graduate education, a three-pronged approach is used that consists of supervisor support, group courses teaching scientific writing, and access to a writing center that provides individualized support in relation to writing particular scientific texts.

At the professional level, scientists still need support for the writing of research articles and other professional duties in English as a second language. Support for these professional scientists should be individualized and directly address their needs in scientific writing in English. The most appropriate model that captures the understandings of scientists in this study is that of a specialized writing center that is open and accessible on an immediate basis and provides high levels of individualized service. These scientists wish to have the opportunity to consult with an expert writer with some degree of scientific knowledge about their professional writing. They do not want a writing center tutor to just correct the manuscript but rather to really interact with the scientists and help them to learn from this interaction. In addition, it is possible that courses or mini-workshops could be provided and designed for scientists, providing insight into the scientific writing process. However, the most useful approach is the provision of ongoing, individualized support in relation to the English scientific writing of these scientists.

It is important to point out that individualized instruction with professional scientists should have quite unique characteristics. These scientists are highly accomplished professionals, and as such, the relationship between the writing tutor and the scientists must be equitable and part of a sharing of expertise. In one sense, this is more of a manifestation of research teamwork, which is very common in science, where an expert is brought in to consult about a specific aspect of the study. Thus, this relationship, and perhaps even aspects of authorship, reflects the nature of the team that actually contributed to the creation of the paper. A further manifestation of this line of thinking includes enhancing the existing writing collaborations situated within natural scientific, collegial relationships. It is possible that a writing tutor could also be consulted so as to provide advice on how best to work with another colleague in relation to issues of co-writing. For example, writing tutors could provide instruction in how to respond to a research article or ways of organizing the co-writing process. The writing tutor, in this sense, is an expert consultant and part of a much broader team of professionals who contribute to the writing of research articles.

The approach outlined by these scientists and summarized here may interact positively with the current situation of enhanced anxiety and fears of the writing of scientific research in English. An extended educational process that develops writing and language skills across higher education in the form of a set of courses and the provision of individualized instruction and support through graduate study and professional life offers a comprehensive solution to the current situation of the imposed requirement that scientific publication be in English. We draw on the results of this study and the published work of others in the next chapter, so that this broad plan of educational intervention is further explicated, detailed, and explained.

8 Facilitating Improved Scientific Writing in English as a Second Language

8.1 Aims and Underpinning Positions

The overall aim of this chapter is to develop an informed approach for assisting second language scientists in their writing of research articles in English as a second language. The research presented in previous chapters quantifies the overall burden involved in writing science in a second language and corroborates existing literature showing that scientists face difficulties in publishing their research in English. For many scientists writing in English as a second language, there is a need for some form of educational intervention and support. This chapter develops our understanding of what can be done from educational and administrative perspectives to provide the necessary educational resources and interventions to allow scientists to overcome the barrier of publishing in English.

As seen in the data presented in this book, scientists writing research articles in a second language are not an undifferentiated, homogenous group. Some senior scientists from the research institute clearly did not need or want any help with their scientific publishing; whereas junior faculty and faculty at the teaching university would clearly benefit from, and desire, educational and administrative interventions to help them with their second language publishing activities. Accordingly, the approach developed here should be able to account for the different needs of different scientists. This is particularly important since the current study focused on scientists in Mexico, but our recommendations address the much wider audience of scientists across the world who write research articles in English as a second language.

A different premise in designing our educational approach is our understanding of the students' involvement in this pedagogy. Our approach is principled upon the notion that our students are capable professionals in their fields of expertise. Some of the beneficiaries of our educational practices may bring to the endeavor of writing scientific articles well-developed literacy skills in their first language, expert reading skills of scientific English, and varying degrees of success in writing scientific articles in a second language. Others may have differing degrees of knowledge on any of these levels. In any case, these students bring with them expert knowledge in their respective fields of science, and this expertise is essentially what the whole educational endeavor of writing research articles attempts to facilitate. Accordingly, an educational approach of this kind with students needs to be based on respect for the scientist and their aim to generate and disseminate new knowledge. As we stated at the very beginning of this book, basically, we believe that the writing of research articles in English should not be a barrier to the generation of new scientific knowledge across the world. The approaches outlined in this chapter are designed to provide ways of overcoming and addressing this potential barrier.

Finally, our broad approach to educational administration, teacher training, and material development is one that is aligned with postmethod approaches that assume that all participants in an educational program are decision makers and problem solvers (Kumaravadivelu, 2002). This is especially true when the audience we deal with includes highly professional and knowledgeable scientists, administrators, and teachers. In other words, as educators and researchers, our desire in this chapter is not to provide a prescriptive and fixed pedagogical solution but rather outline the issues and specify types of educational interventions that are possible. Additionally, we emphasize that the particular educational programs that are initiated must be based in the context of the scientific population, institutional resources and academic/cultural contexts specific to each locale. Based on the research we collected from scientists and the larger literature concerned with specific support, we identify what can be done and the organization of such a program of education. We also offer modularity in designing programs in different ways because we recognize that decisions will have to be made at every educational site. Readers of this book will want to determine what specific interventions fit best within their context.

To address the diversified nature of scientists writing research articles across the world, the importance we assign to both the scientist and the development of scientific knowledge and our desire to facilitate diversified educational programs, this chapter is organized to provide a modular, problem solving approach to the development of research article writing abilities. The beginning section of the chapter presents a series of core principles based on our research and a review of knowledge in the field. These principles provide an informed understanding of the issues involved in facilitating scientific research article writing in English as a second language and construct a heuristic frame within which other educators, scientists, and administrators can consider their own educational programs. The second section of the chapter presents a series of educational interventions that can be used to construct and facilitate scientists in writing research articles in English. These educational interventions are the building blocks of an educational program. The third section of this chapter outlines our suggestions for the design of an extended educational program for facilitating research article writing based on the data we collected from our sample of scientists in a Mexican context.

8.2 Principles and Recommendations

Based on our research and review of the existing literature, the following principles and recommendations provide a set of parameters within which programs can be designed to address the needs of different scientists contending with the task of writing research articles in English.

8.2.1 Long Term Commitment to Writing Education

Developing the ability to write a research article in English as a second language involves an extended educational process. It is clear from what we know in general about literacy education, and specifically about the development of scientific writing in a second language, that learning to write takes a long time (Bazerman, 2007). Based on the data presented in this book, the recommendation is to develop writing from undergraduate studies all the way through doctoral studies and provide additional support well into graduates' professional lives as scientists. Thus, the first core principle is the understanding that facilitating the process of writing research articles in English as a second language needs to span the full length of higher education and professional work (Duff, 2010; McGinn & Roth, 1999).

8.2.2 Differential Needs and Diversified Educational Interventions

Scientists who write research articles in a second language are not a homogenous group, and have different educational needs and desires. Accordingly, in developing an approach to the facilitation of research article writing, a range of approaches needs to be employed. The principle is that a diversified set of educational interventions, addressing the different needs of scientists at various stages of their development (e.g. Cargill, 2011) and with differing backgrounds, will need different types of support and instruction. In designing long term educational approaches of this sort, educators and administrators need to provide a range of responses and not rely on a single mode of educational intervention, such as a compartmentalized course on writing, as the standard method of addressing research article writing. The recommendation is that long term educational programs designed to enhance and facilitate scientific research writing in English as a second language provide a range of interventions of different types based on the differential needs of the scientists.

8.2.3 Multilayered Understanding of the Research Article

A scientific research article is a complex, socially embedded literacy product. As seen in the review of the research article and the evidence presented in this book, writing a research article involves structural linguistic features on the micro and macro levels, psychological process of production, and social contextualization in a range of settings and research networks. Accordingly, educational interventions of different types need to address the complexity of the research article and offer support for the various aspects of writing a research article. It is a mistake to consider writing instruction only in relation to the linguistic features of the research article (Belcher & Conner, 2001; Ivanic, 1998). The recommendation is that educational interventions address the whole range of factors that are functional in facilitating the writing of a research article. The principle is that writing instruction for research articles address the multiple layers involved in producing and writing a research article in English as a second language.

8.2.4 Provision of Expert Support for Science and Writing

Writing a scientific research article involves the integration of scientific and literacy knowledge. Accordingly, in designing educational interventions to facilitate the process of writing a scientific research article, support needs to be provided by experts in science and literacy (e.g., Cargill, O'Connor, & Li, 2012; Poe, Lerner, & Craig, 2010). This means that the issue of developing the ability to write a research article needs to involve senior scientists and researchers working with other scientists who may need help and support in developing their abilities to conceptualize and express their scientific arguments. Expert literacy instruction and support is also required to explain the structures and processes of research article writing and offer continuing support for the writing process (Kennedy, 1997). The recommendation is that educational programs provide access to both types of expertise. The principle is that learning to write a research article involves developing skills and knowledge in relation to science and the literacy of science.

8.2.5 Personalized, Continual, and Immediate Support for Research Article Writing

As reported by our participants, support for writing a research article involves the ability to be able to contact an expert about specific and contextualized problems that arise during the actual process of writing. Furthermore, as seen in our data, different scientists require help with various aspects of writing a research article. This means that personalized, individualized support is needed for scientists in a framework that allows them immediate and continual access when and where they need it. The principle is that scientists have easy access to expert writing knowledge during the writing process itself. A writing center exemplifies this type of writing instruction in which tutors work one-on-one with students who require support for their different writing assignments. The recommendation is for a support center of this type that provides personalized instruction and support for the writing of research articles as they are written by the scientists be provided.

8.2.6 Demystification of the Structures and Processes of Scientific Publication

As reported by our participants, there is the belief that explicit instruction in the features and processes of scientific writing would be beneficial. There is a need to clearly explain and explicitly teach the relevant aspects of the research article, and also the psychological and sociological processes involved in getting a research article published (Englander, 2009; Lillis & Curry, 2010). Basically, the process of scientific publication needs to be demystified to allow the maximum amount of control and understanding of this process to be provided for scientists. The principle is that scientists need to be taught explicitly about the research article and the psychological and sociological processes involved. The recommendation is that courses, workshops, and presentations incorporate this explicit instruction.

8.2.7 Broad Administrative, Institutional, and Financial Support

Overcoming the barriers of scientific publication in English as a second language requires broad administrative support and dedicated funding. To really address this issue and provide much broader access to scientists around the world, administrators and institutions need to be prepared to commit time and resources to the construction of an extended program that supports this endeavor. The basic principle here is that being able to write a research article in ESL is not only the problem of the individual scientist but, rather, a structural, administrative and institutional issue requiring all interested bodies to provide support and solutions. The tendency may have been to compartmentalize the problem and provide a writing course or a writing instructor as a solution. But, the data presented in this book suggests that a much broader approach must be taken that integrates writing courses, personalized support, access to research networks, and mentorship from established scientists. This long term and diversified set of interventions requires dedicated planning, support, and budgetary resources.

8.3 EDUCATIONAL INTERVENTIONS

As described above, designing an educational process that can facilitate the development of scientific research article writing in English as a second language is an extended process involving a range of different educational interventions that address the needs of various scientists. In this section of the chapter, the types of educational interventions are outlined. The aim is to delineate a wide range of types of interventions, and in each case, explain the objectives, types of participants, methods of functioning, and limitations. In this way, the building blocks of an educational design for facilitating research article writing in English as a second language can be delineated. We emphasize that the particular configuration in any one institution is determined by many contextual factors. The funders, administrators, and faculty involved need to situate the educational programs detailed below in a manner that is most appropriate for their context.

8.3.1 Explicit Teaching

Writing scientific articles in English requires knowing about language, rhetoric, publishing, and networks. This knowledge is necessary in addition to expertise in the scientific discipline and having data to contribute to it. Research conducted in the fields of applied linguistics, sociology of scientific knowledge, and rhetoric and composition about writing scientific articles can be made explicit through courses and workshops for scientists and students. In this section, we describe how explicit teaching can be conducted and what topics might fruitfully be presented.

Description. Explicit teaching presents a structured set of content and activities for the benefit of the participants. The teacher is expected to be an expert in the content areas. There are many possible pedagogical designs for the delivery of the content, from one-time presentations to online distance education, to formal, credit-earning courses. The literature generally discourages one-time lectures and reports that more sustained programs are likely to have a greater impact on improving faculty research article production (McGrail, Rickard, & Jones, 2006).

Purpose. The purpose of explicitly teaching scientific research writing is to demystify the process of publishing and provide the participants with the skills to participate successfully as authors. The structured delivery of the content allows the participants to integrate the new knowledge into their existing schema. The structured delivery is accompanied by activities that the participants perform individually, or cooperatively with others, to practice and reinforce learning.

Participants. Participants in explicit teaching are self-selected based on the content of the course or workshop. Depending on the institutional configuration of the teaching, they may earn certificates, credit hours, professional development recognition, or other acknowledgements for their participation.

Method. Explicit teaching, as presented here, has a set of three principal, content modules with subsections (see Figure 8.1). The principal modules are "The Scientific Publishing World," "English Research Article Structure," and "Writing Troublespots." The subsections of each module develop different aspects of the topic. Participants could enroll in the entire module, or depending on the institutional configuration, pick and choose the subsections from across any of the modules.

Two supplementary modules might also be developed although they are not directly related to writing science articles. One supplementary module could focus specifically on "Reading and Comprehending Science Articles in English" while another could focus on "Effective Conference Presentations" in English or the participants' first language.

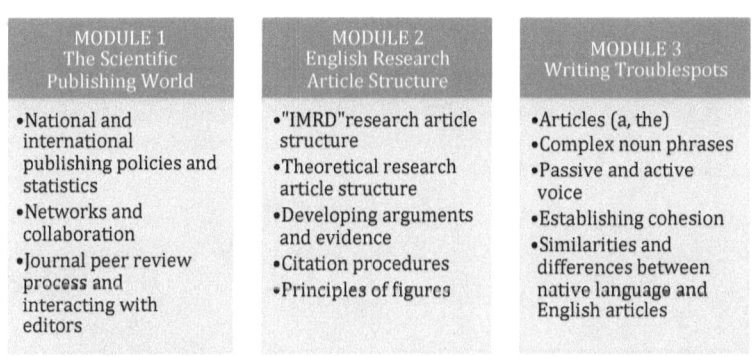

Figure 8.1 Three explicit teaching modules and related subsections.

The delivery of the content can vary along several dimensions. The host institution can structure the explicit teaching activities using the "Content Delivery Matrix" to identify the most appropriate configuration (Figure 8.2). The relevant dimensions that determine the structure are configuration, mode, duration, evaluation, recognition, and participant investment.

Context Delivery Matrix Variable Elements	Options for delivering the variable elements			
Configuration	Complete module with subsections	Independent subsections		
Mode	Small group face-to-face	Large lecture face-to-face	Synchronous online	Asynchronous online
Duration	One-time presentation	One/two-day workshop	Several week course	Semester course
Evaluation	Participation only	Pass/fail	Graded	
Recognition	Certificate for all or part	Release time	Professional development recognition	Academic credit
Participant investment	Fee charged to participants	No fee to participants		

Figure 8.2 Explicit teaching Content Delivery Matrix.

Advantages and Limitations. Explicit teaching is a learning mode that is familiar to teachers and students. The content can be structured in ways that facilitate individual and group activities. The Content Delivery Matrix demonstrates a range of possible configurations within the institutional framework. Multiple variations are possible as well. It is worth noting that there is a growing array of books commercially available to support this content delivery. Recent titles include *Writing Scientific Research Articles: Strategy and Steps* (Cargill & O'Connor, 2009); *Writing Your Journal Article in 12 Weeks* (Belcher, 2009); *Academic Writing and Publishing: A Practical Handbook* (Hartley, 2008); *Getting Published in International Journals* (Reid,

2010); *Writing for Academic Journals* (Murray, 2009); *From Research to Manuscript* (Katz, 2009); and *Science Research Writing: A Guide for Non-Native Speakers of English* (Glasman-Deal, 2010). Mary Jane Curry (2011) recently published a review of several of the titles. The combination of familiar teaching-learning expectations, varied delivery options, and existing teaching materials are the principle advantages of explicit teaching.

The literature reports on a number of cases of explicit teaching in foreign language contexts. For example, Cargill, O'Connor, and Li (2012) discuss their Collaborative Interdisciplinary Publication Skills Education program. Moreno (2010) discusses a pedagogical approach focusing on cross-cultural variation in academic writing in English and Spanish. Storch and Tapper (2009) quantitatively evaluate the impact of a semester-long English for academic purposes course with international postgraduate students in Australia. Improvements were measured in grammatical accuracy, text structure, and rhetorical quality. Other experiences were more nuanced. For example, Craig, Poe, and González-Rojas (2010) report on the challenges of implementing an explicit, interdisciplinary approach to writing in two universities in Mexico. Assumptions on the part of the American and the Mexican collaborators were sometimes at odds, leading to difficulties. A course designed to improve presentation skills, rather than writing, for English as a second language scientists is outlined by Teresa Morrell Moll (2009). In an alternative model, American universities have also begun to offer intensive, on-site writing courses for international scholars. For example, in 2009, Texas A&M University began offering an annual, three-week summer course. At the end of the course, scientists are expected to come away with a manuscript ready for journal submission. Such intensive programs may be useful, but they are expensive ($1,500 before food and lodging) and require an extended stay away from one's other responsibilities.

A common element of explicit teaching is that, typically, it is not individualized. Thus, a limitation of this form of support to scientists is that their personal writing needs may not be directly met. The format of explicit teaching may require an ongoing commitment of time inside and outside of the classroom. Active scientists may be unwilling or unable to make such a commitment. Thus, the limitations of this form of support for professional scientists are based on the degree to which the content is generalized and not individualized, and the activities are time-consuming beyond the participant's existing commitments.

8.3.2 Collaborative, Face-To-Face Interaction with Associates

The model of collaborative, face-to-face interaction is an individualized manner of support to an author. The collaborator is called an associate in this model. Rather than working in a group setting, the author and the associate meet together to focus on assisting the author to present the best possible text. The focus could be on preparing a written text such as a conference abstract, grant proposal, or research article. On the other hand, the focus could be on oral texts, such as preparing and practicing a conference presentation. This model provides individualized assistance to the author.

The institution should be a specialized location for this service. This location needs to be quiet and comfortable. The interactions between author and associate need to be private because the author's content is likely to be confidential and proprietary. The physical location must be designed to safeguard this confidentiality, and the physical space needs to provide access to resources such as databases, corpuses, and full-text journals.

Description. Collaborative, face-to-face interaction is based on a relationship in which different bodies of knowledge held by the participants are integrated so as to facilitate completion of the writing task. The two participants come together intending to support the author's creation of a specific text that is intended for a specific venue. The author is the expert on the subject matter of the text, that is, the research to be presented. The associate brings English-language linguistic knowledge or subject area expertise in scientific writing in the disciplines. The two participants meet in an ongoing relationship to accomplish the specific writing or presentation project.

Scientific communication is not a singular activity and may involve many iterations of the same document. Scientific communication can also comprise many different documents, from conference abstracts to complete book manuscripts. Therefore, the associations established in this one-to-one relationship are likely to extend to meetings over numerous projects.

Purpose. The purpose of collaborative, face-to-face interaction is to support the author in accomplishing particular English-language science communication projects. The collaborative nature of this model makes it different from more traditional writing tutor models (Murphy & Sherwood, 2011). Here, we presuppose that each member of the collaborative relationship will expand their knowledge base. In this arrangement, the author who seeks support also serves as an informant to the associate about scientific research. As a consequence, the relationship becomes an enriching one for each member of the association. The aim is to develop increasing autonomy in the author's English-medium science activities.

Participants. The author-participant may be a scientist or student who seeks assistance with a scientific communication task. It is assumed that the author wants to broaden their knowledge base to develop increasing expertise and autonomy.

The associates in this model bring two different kinds of expertise, and so the associates are likely to be two different individuals. One type of associate is a person with English-language, academic writing expertise. This is a person knowledgeable of the lexical-syntactic system of English and who is able to be explicit with that knowledge. This knowledge also includes understanding English-language expectations regarding scientific genre conventions and rhetorical structures. The other type of associate is a person with scientific disciplinary expertise. This is a person who can work with the author in distilling the most salient findings, refining the scientific claim, and positioning the research in the international conversations of the field. In other words, this is a person who understands the disciplinary demands of a piece of research and assists the author in formulating a fundamental knowledge contribution. The author may seek both types of associates, depending on the immediate project. For example, writing a response to a journal reviewer may involve both disciplinary positioning and linguistic appropriateness. It is ideal for the associates to be bilingual.

Method. Collaborative, face-to-face support requires that the associates receive training. The training is modeled, to a large extent, on the North American writing center tutoring relationship. In this model, the associate/tutor is trained to pose questions to the author to identify the author's needs and the shortcomings of the immediate text.

The author is guided through the processes necessary to improve the text. It is presumed that the knowledge gained in this interaction is transferrable to future writing tasks. Thus, the method involves exposing the author to knowledge and skills that are broadly applicable for future writing projects.

An important consideration in this model of support is the notion of collaboration. The language or science writing expert is also expected to learn from the author. The author will expose the associates to the research conducted. This broadens the associates' knowledge as well, and it creates a relationship of true interchange. Interchange is necessary in supporting the equal but different fields of knowledge that reinforces the strengths of each member of the collaboration.

As noted in the "Participants" section above, two types of associates are needed for this support to be wholly effective. Each type of associate brings specialized knowledge. The specialized knowledge is useful to the author for different kinds of concerns. As shown in Figure 8.3, the author seeks out one type of associate or the other, depending on the immediate concerns.

Advantages and Limitations. The advantage of collaborative, face-to-face interactions is that they are personalized to the needs of the author. The author has a one-to-one relationship with the associate and is able to call upon the associate's specialized knowledge as needed. There are two types of knowledge available to the author, although it is recognized that these are not wholly independent of one another. This model also offers ongoing support so that one project may be completed over several meetings, or so that several projects can be completed. It is intended that the author will develop increasing autonomy over time. This model is also premised on true collaboration, such that the author also supports the associate with greater scientific knowledge based on the particular research project the author has undertaken.

This model requires that specialized associates be trained and available to work with authors. The nature of scientific publishing requires the associates to be virtually on-call in order to comply with the strict deadlines the authors face. This program of support needs to be housed in a comfortable and accessible location that provides privacy and confidentiality for the authors.

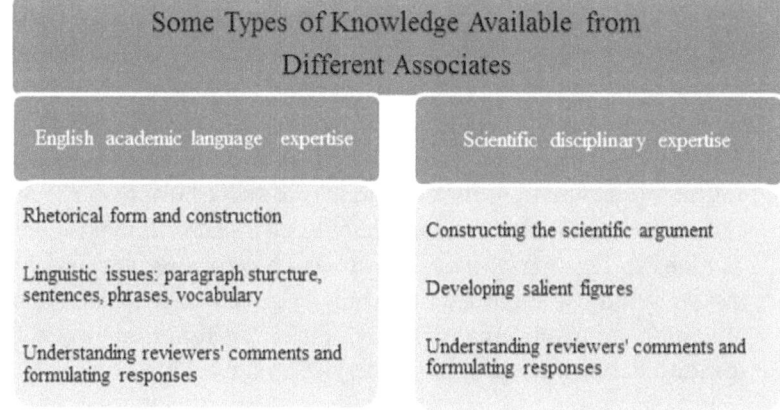

Figure 8.3 Types of knowledge that different types of associates bring to face-to-face interactions.

8.3.3 Expert and Peer Collaborations

Science is rarely performed alone, and writing up scientific research is rarely done in isolation. The collaboration which is typical among colleagues and between supervisors and graduate students is formalized in two program models for professional scientists. These program models build on the successful writing experiences that scientists report in their trajectory to becoming successful writers of scientific articles when English is not their first language.

8.3.3.1 Mentoring Program

Description. A mentoring program matches a novice member of the field with an experienced member. Mentoring is typically an informal association between individuals; however, this mentoring program model is intended to be formalized through institutional support and recognition.

Purpose. The purpose of mentoring is to provide the novice with "insider" knowledge unlikely to be available due to one's novice status (Nakamura, Shernoff, & Hooker, 2009). Experts, or established scientists, have a great deal of insider knowledge about the relationships and networks that facilitate access to international research and publishing. The range of insider knowledge is vast. Insiders' information that facilitates publishing includes where to publish, the preferences and idiosyncrasies of individual journal editors and editorial boards, knowledge of upcoming special issues and "hot topics," and guidance on responding to journal reviews. Other insider information that assists novice scholars includes sharing knowledge of which conferences are important, making introductions to other scholars who may facilitate international and/or prestigious collaborations, making introductions to funding officers from research programs, and providing insight on how to build a solid, scholarly career. In addition, mentors offer highly valuable insights to novice scientists regarding negotiating institutional issues of promotion, service, and teaching load.

Participants. The mentoring model is founded on the relationship between a senior scientist and a novice scientist. This may be a one-to-one relationship, or one senior scientist may mentor a few junior colleagues. The scientists work within the same institution and in similar, if not identical, sub-disciplines of their field.

Method. The method proposed here is based on the knowledge that the mentor offers an invaluable service to the novice; therefore, institutional recognition for the mentor's service is crucial (Kennedy, 1997). Initial training must be available to mentors so they are aware of the scope of their role. A mentor's service should be recognized through release time or with additional resources. As an institutionalized program model, the mentor and novice should be granted resources for meeting and discussing relevant content, on at least a monthly basis, throughout the novice's ascent to a tenured position, and continue for at least one year beyond that milestone. It is commonly acknowledged that establishing oneself in a scientific field takes from six to ten years beyond the completion of a doctorate.

Each new scientist who joins the institution is matched with a mentor. The meetings between expert and novice do not have a formal agenda. However, both members of this relationship come to each meeting with information to share. The topics of discussion focus on professional opportunities and trajectories, as mentioned in the "Purpose" section above. As the novice becomes established, he or she becomes a mentor to a novice entering the institution.

Advantages and Limitations. A mentoring relationship is built on good intentions. If the individuals are not suited to one another, a new pairing of mentor and novice may be necessary. On the other hand, a well-suited relationship may continue informally, long after the initial, formal period of approximately six years. It is not unusual in the sciences for even senior people to consult with those who served a mentoring function early in their careers. Therefore, an institutionalized mentor program model will help to assure that novices are quickly enculturated into their disciplinary practices and their ability to succeed is facilitated. Apart from some institutional recognition provided to the mentors, a mentoring program model is inexpensive for an institution to initiate and maintain. There are few direct costs, and no expectation of hiring additional staff or expertise once the initial model is put into place.

8.3.3.2 Faculty Writing Circle

Description. A faculty writing circle is a regular meeting of faculty who gather to read and comment on the written work of members of the group (Rankin, 2001; Weissberg, 2007). This intervention develops peer feedback skills for faculty members. Formal membership in the group is voluntary, but a commitment to participate for a specified period of time, such as four, six, or twelve months, is necessary. A well-constituted faculty writing circle can continue, literally, for years.

Purpose. The purpose of a faculty writing circle is to provide constructive feedback to the faculty member on a manuscript intended for publication in a context of mutual respect and support (Rankin, 2001). Having a manuscript read by peers prior to submission is useful for the author to determine its readiness. Peers provide feedback on multiple aspects of the text even though they are not necessarily disciplinary experts in the subject matter of the manuscript. The members of the writing circle can provide feedback on issues of, for example, manuscript structure, development, linguistic construction, and figures. The author can then make whatever revisions seem appropriate in order to further assure acceptance of the manuscript by the target journal.

A crucially important aspect of a faculty writing circle is that it can support the affective side of scientific writing. Issues of anxiety and risk that can occur in the publishing process can be discussed and addressed in a context that is mutually supportive and respectful.

Participants. The participants are faculty members within the institution. The number of participants ranges from at least three and up to about nine. This number assures that the author receives feedback from more than one person, and also that all feedback can be incorporated in one meeting. If there are many more than nine participants, the opportunity for each author's work to be read is lessened because only one manuscript is read at each meeting of the faculty writing circle. If there are nine members, a cycle of nine meetings is necessary for everyone's work to be read at least one time.

Typically, the members are not part of the same research group, nor do they typically collaborate as authors in a project. In fact, they are not necessarily situated in the same discipline or subdiscipline. Multidisciplinary circles can be particularly enriching because disciplinary epistemologies and conventions are explicitly explored when there is variety.

Because the purpose of the faculty writing circle is to enhance the acceptance of manuscripts by journal editors, issues of English-language appropriateness and accuracy may be involved. None of the participants is expected to be a native English speaker, but their knowledge of English science writing based on their experiences with reading and writing scientific texts in English is valued. Feedback regarding vocabulary, syntax, and genre, in addition to rhetorical development, is to be provided to the author.

Method: Faculty writing circles are flexible in structure and method, but there are common traits. Generally speaking, they establish a regular meeting time of once a week, once every two weeks, or once a month. They meet for about 90 minutes in a location that is comfortable on or off the institution's grounds. At each meeting, a manuscript written by a group member is discussed.

To facilitate the most productive use of the meeting time, a procedure for the discussion has been developed (Figure 8.4). Basically, the author sends the manuscript to the writing circle members one week before the meeting with elementary information. This information includes the genre of the manuscript, audience, stage of development, and the author's particular concerns. The members read the manuscript and prepare to comment in the meeting.

At the meeting, all the members of the writing circle focus on the manuscript. If it is possible to display the manuscript on a projector, this is ideal so that all the members can refer to the same section or sections without confusion. The meeting begins with the author offering a very brief summary of the manuscript. Then, comments from members proceed in a series of four "rounds" in which different kinds of information are offered to the author. The first round focuses on clarification; round two offers positive comments; round three responds to the author's initial questions; and finally, further comments and suggestions are discussed. The author is encouraged to listen and take notes but not reply. The author is discouraged from being required or expected to justify or explain. The author is encouraged to take advantage of the well-intentioned comments from the other writing circle members. Ultimately, it is entirely the decision of the author as to what revisions, if any, are made to the manuscript. The author may or may not wish for the writing circle to review the same manuscript at a later date. Before the conclusion of each meeting, the group should decide whose manuscript is to be read next.

Faculty Writing Circle Procedure

Writer's Responsibilities: One week before the week we read your manuscript, you will

1. Send by email and attachment your article or portion of your article that you want us to read.
2. Attach a cover sheet in which you explain to us:
    a) What the piece of writing is (for example, a full article, a 'letter,' memorias, the methods and results sections of an article, etc.)
    b) What audience is it intended for (is it a national/international journal? non-refereed magazine?)
    c) What draft stage is it in (first draft, major revised, final?)
    d) What particular questions you would like us to address. (For example, Is the discussion well related to the introduction? Is the Method section clear? Is the transition between paragraph 3 and 4 in the conclusion good? Am I respectful of other authors whom I am criticizing?, etc.)

Remember, we do not know your field nor what is typical in your field's research and publishing. But, we are all experienced readers of scientific research and we all want to improve our writing abilities.

Reader's Responsibilities

1. Take the time to read carefully and generously. We are helping the writer to prepare his/her manuscript for its particular audience, so read the cover sheet that accompanies the manuscript.
2. Keep in mind the stage of the manuscript (step 2a above). If it is a first draft, do not worry too much about sentence level errors; if it is ready to be sent to a journal next week, do not suggest a whole new approach. (Suggestions for different levels of response are included below.)

Workshop Procedure

We will have four 'rounds' or discussions about the writer's text. It is the writer's job to mostly listen and take notes. The writer may want to explain, argue, or defend your choices rather than simply absorb responses. It is your work and you are the final arbiter. It's best to listen rather than talk.

Round 1: Clarification. The readers ask general questions if necessary to clarify, for example, intended audience, certain key terms.

Round 2: Positive Comments. Each reader comments on one thing they specifically liked or admired. You might comment on the organization, content, voice, style, any aspect of the writing that you want to emphasize. Be relatively specific! Just saying, "I liked it, it's good" doesn't help the writer to understand his/her strengths.

Round 3: Answer Writer's Questions. The writer sent us a cover sheet with specific questions she/he wanted us to focus on. We will now answer those questions, even if we think there are other things to talk about. (Remember, the writer is listening and taking notes, not defending or justifying the writing.)

Round 4: Other Comments, Questions and Suggestions. We can mention other things we think are important in the manuscript. We will keep a positive tone and even if we are critical we want to say it so that it is helpful feedback.

If we made notes on the manuscript, we can give this to the writer if appropriate.

The writer leaves the meeting with information that we offer to help him or her improve the manuscript. The writer is not obligated to revise the manuscript, nor show us the revisions. Of course, the writer may show us the revisions at a later session if he/she wants to.

Levels for Response for Reading Drafts*

1. Large-scale concerns.

- Is the purpose of the manuscript clear?
- Does it seem to be appropriate for its intended audience?
- Are the ideas explained carefully enough?
- Is the form consistent with the expected academic/professional guidelines?
- Are the overall voice and tone appropriate (not too formal nor too casual)?

2. Mid-level concerns.

- Is the order of ideas logical? Does anything seem to be out of place?
- Are the sections and paragraph breaks logical and appropriate?
- Are the transitions between sections or ideas smooth?
- Are there enough signals and subheadings to help the reader?

3. Sentence-level concerns.

- Are the sentences clear and mechanically correct?
- Are there any misused words or awkward phrases?
- Is the punctuation appropriate and correct?
- Could the language be made more concise or precise?
- Could the language be made more active and lively?

*Source: Rankin, E. (2001). The work of writing: Insights and strategies for academics and professionals. San Francisco: Jossey-Bass. p. 100. Used with permission of John Wiley & Sons, Inc.

Figure 8.4 Suggested procedure for Faculty Writing Circle meetings.

Advantages and Limitations. Faculty writing circles are inexpensive for an institution to support because no additional resources are necessary beyond the time invested by the individual members. Faculty writing circles are flexible in the range of genres that the members discuss. Any item in the entire "genre chain" (Swales, 2004) of professional scientific writing could be focused on by the group, including conference proposals and abstracts, conference presentation materials, papers for journals or proceedings, and responses to journal editors. A faculty writing circle can be constituted for a limited period of time, such as one semester, or it may continue almost indefinitely according to the wishes of its members. The format of a writing circle is sufficiently flexible in that alternate configurations are possible. For example, the writing circle could meet in a retreat setting, where all the members' manuscripts are read during a condensed period, such as a weekend. The multidisciplinary representation of the members can lead to a greater understanding of one's own discipline as discussions invoke different genre formats and epistemological assumptions. This varied representation could possibly lead to interesting interdisciplinary collaborations as well.

The greatest limitation of faculty writing circles is centered on the interpersonal skills of the members. Because this is an activity without a clear leader or outside expert, the dynamics of the circle are dependent upon the goodwill and respectful collaboration of all the members. Long term success of a faculty writing circle evolves with the mutual commitment of members.

8.3.4 Translation and Editing Services

Scientists rely on their first language in different ways when engaged in scientific writing and communication. Some use their first language for the cognitive-intellectual work of conceptualizing communication tasks and content; some use their first language to develop an outline or draft of their text; some entirely write their manuscript in their first language. As the research of this book attests, not all scientists are sufficiently knowledgeable of English to produce a manuscript that is in a form suitable for submission to an English-medium journal. Therefore, services to render a manuscript in English are necessary.

Description. Translation and editing services are related but separate activities available to the second language, English-speaking authors. A translation service takes a manuscript produced in one language and reproduces it in another. An editing service takes a manuscript produced in English by someone who is not a first language speaker of English, and revises it for linguistic and rhetorical acceptability.

Purpose. The purpose of translation and editing is to provide second language, English-speaking authors with services that their language skills are insufficient to produce. These services are based on the premise that producing "linguistically adequate texts" when the scientist is a second language speaker requires a level of investment of time and effort that can be burdensome (Ammon, 2001, p. vii). Despite making such an investment, the scientist still may not achieve the necessary proficiency for their manuscripts to be accepted by journal editors without language criticism. Therefore, these services are provided to lessen the English-language burden placed upon scientists and still permit the scientist to contribute to the knowledge production of their discipline.

Participants. The two services of translation and editing could be performed by people with the same profile, but this is not necessary. Translators need to be fully bilingual in both the native language and English to render the original text faithfully into English. Translators should have at least some basic scientific disciplinary knowledge. Editors need to be expert in English-language skills, but their knowledge of the native language is less crucial in their role. The editor works with a text that is already in English. The editor must be able to follow the rhetorical arguments of the text and work with the author to clarify issues of ambiguity or confusion. People providing such services are also termed "literacy brokers" in that they shape the manuscript the scientist produces (Lillis & Curry, 2006).

Method. Translation and editing services function as a drop-off center for the author and the manuscript. The author indicates the type of service needed, either full translation or editing. The turnaround time for completion of the service is negotiated based on the author, the deadline the author faces, and the workload of the service provider. The author needs to be available to clarify portions of the manuscript that may be unclear, ambiguous, or confusing to the translator or editor.

Rhetorically, the genre of a scientific article in the first language may differ from the genre of a scientific article in English (see Chapter 3 for a fuller discussion of these issues). The translators and editors must be knowledgeable of these varying conventions and prepared to discuss them with the author. The ethics of translation services must be explicit in the service provided to the author.

An additional service offered is phone-in or drop-in quick translation of phrases and expressions. Authors who are otherwise adept at writing in English may have an immediate, specific need for translation of a word or phrase. This service should be available as well.

Advantages and Limitations. Translation and editing services provide an efficient means for second language, English-speaking scientists to publish in English-medium journals without making the enormous investment to learn to produce linguistically adequate texts. Translation allows a scientist to produce a manuscript entirely in their first language. Editing services allows the scientist who has produced an English-language text to receive a review and improve the rhetorical quality of the manuscript before submitting it to a journal. These services facilitate the smooth and timely acceptance of a manuscript in a field where timeliness is important to make first knowledge claims.

The limitation of these services is first that the author must be able to review the manuscript to assure that the original knowledge and intent is rendered accurately in English. Because scientists read predominantly in English in their discipline, it is expected that the scientist is able to determine whether the English manuscript is true to the original intent. Additional time may be required for subsequent revision by translators or editors to comply with the author's scientific and rhetorical intentions. Another limitation of translation and editing services is that the author's skill as a writer in English may not improve because the author relies on the services of a translator or editor. These translation and editing services remain crucially important; however, because the expectation that all scientists will achieve the necessary proficiency to write scientific articles successfully in English is unreasonable.

8.4 Educational Program

In this section, an extended educational program designed to alleviate the burden of writing a research article in English as a second language is outlined. The model of educational intervention is based on the research reported in this book and the principles and instructional interventions outlined in the previous section. As a basic principle, we consider the educational process to extend from undergraduate study through professional work as a scientist. Figure 8.5 presents a schematic view of the whole educational program.

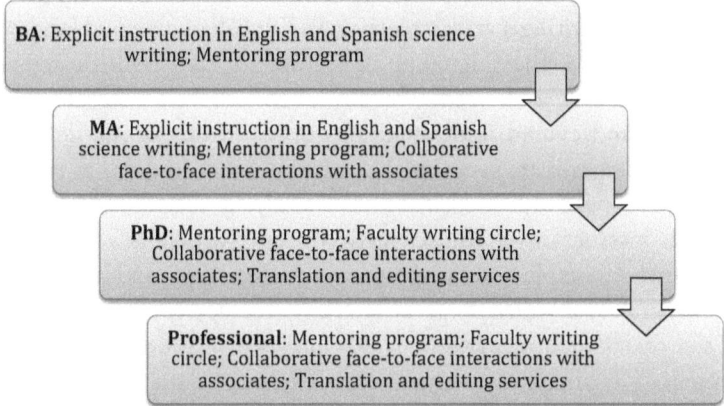

Figure 8.5 Development of educational program

As can be seen in Figure 8.5, the educational program starts at the undergraduate level and continues through to professional life. The overall direction of development is from explicit group instruction to individual and personalized instruction and support. The coming section specifies the proposed educational interventions at each different educational level.

Bachelor's level. At the undergraduate level, knowledge of scientific writing needs to be developed through explicit instruction. These courses should relate to scientific writing in both Spanish (first language) and English as a second language. Furthermore, the full range of scientific writing issues should be addressed, including micro and macro level linguistic features and an understanding of the world of scientific publishing. This development of explicit knowledge of research writing should be complemented through interaction with professional scientists and through access to professional publication of scientific research. These mentorship activities could be as simple as a discussion concerning the scientist's research agenda and publishing plans, or as advanced as providing drafts of research articles and discussing them. The important aspect of this mentoring program is that undergraduate students get the opportunity to connect the writing practices of science with the actual work of a publishing scientist.

Master's level. At the master's level, there is still a need for explicit instruction on writing a research article in English as a second language, and once again, the full range of features of the scientific research article should be addressed. The aim of this explicit instruction is to reinforce and extend understandings of scientific writing developed in the undergraduate program. It is assumed that students at the master's level have committed to actually studying science, and that this explicit instruction of the scientific research article is a major aim. Furthermore, explicit writing courses of this type and at this level could address a wider range of scientific literacy genres, such as the poster or oral lecture, providing additional support for the professional presentation of scientific knowledge. The role of the mentoring program becomes particularly important at this level, since interaction with more seasoned researchers is one main way a young scientist can learn the literacy practices and scientific thinking and the procedures of a professional scientist. It is further assumed that at the master's level, and particularly towards the end of the master's when a thesis is being written, that the student will benefit from individualized instruction in the form of collaborative, face-to-face interaction with a writing expert. This collaborative support is designed to help with actually producing the written components a master's thesis and its subsequent publication as a research article. Thus, the master's student is provided with two different sources of individualized support: from a seasoned scientist in relation to science, and writing a thesis and from an associate who specializes in scientific writing.

Doctoral. At the doctoral level, it is assumed that a young scientist will benefit from a range of different individualized, professional educational interventions. Obviously, the doctoral student needs to be mentored in relation to her/his doctorate from an experienced scientific advisor. This advice should explicitly address the writing of scientific research and not only the scientific content of this work. To provide additional support, the doctoral student should have access to an associate writing expert to work closely with the student on different publishing assignments that they are involved with. This can include help in relation to research articles, posters, and scientific lectures. As specified in the last section, this should be a collaborative interaction in which the doctoral student learns literacy through the process of working collaboratively with an informed writing expert. To further enhance this process, a doctoral student should be directly involved in faculty writing circles as an active participant. These regular meetings around the writing of professional research articles provide opportunities for learning about the real issues faced by scientists in their communities. Doctoral students in these circles should be encouraged to participate by bringing their own writing to these settings and getting feedback from the whole group. Finally, if necessary, doctoral students should have access to translation and editorial services to polish any research articles sent out for publication.

Professional Scientists. As a basic principle, it is assumed that learning to write a fully polished, professional research article does not end after completing a dissertation but is a process that should continue into the professional life of a scientist. As such, the same instructional interventions that characterize literacy instruction during the dissertation process should continue into professional life. First, a mentor who is a seasoned scientist accustomed to the publication process is very helpful to scientists who are struggling to produce research articles in English as a second language. It is assumed that this mentorship will address both the science and the literacy of research publication. Furthermore, professional scientists benefit from extended access to collaborative, face-to-face interactions with writing associates. This form of intervention is crucial at this stage because it offers the chance to both develop professional research articles and simultaneously receive individualized instruction directly on the issues of concern to the scientist. Faculty writing circles provide an opportunity to get feedback on current research writing and to learn from the experiences of others. This should be a routine aspect of the scientist's writing. Finally, if the scientist is overwhelmed with the process of writing up research in English as a second language, translation and editing services should be provided. Translation services should be used in conjunction with writing circles and mentors so as to make sure through collaborative interaction that the translated article fulfills the researcher's scientific aims. Editing services provide a safety net to ensure that final written articles are on the linguistic level they need to be to get published. Together, this collection of educational resources offer the professional scientist different routes for supporting her/his publication activity.

9 Practical and Policy Implications of Supporting Second Language Scientific Writing

9.1 Introduction

In the contemporary globalized world, scientific research is overwhelmingly reported and disseminated in English. In this book, we have addressed the issues that impact science and scientists when English is not a scientist's native language but a second language. We have also examined the challenges that non-Anglophone scientists face when they are physically situated outside the Anglophone center countries of the U.S., U.K., Canada, Australia, and New Zealand. The discussions in this book are informed by empirical research conducted over the past two decades and a recent study involving 150 scientists in Mexico. It is obvious that there are many issues affecting the ability and efficacy of international scientists in their publishing success. These issues stem from individual competence in the foreign language, English, their professional and emotional experiences with publishing, their educational trajectories, their networking opportunities, and the institutional support they receive.

In this final chapter, we outline the practical and policy implications of substantially improving access to English-medium publishing on the part of international scientists. To contextualize this final discussion, we review the principal findings of the research presented in this book. These findings make it clear that participating in English-language publishing presents a burden that is distinctly different from publishing in one's native language. Therefore, we proceed with the knowledge that the challenges that English as a second language scientists face in writing are over and above the challenges of native English-speaking scientists. The findings are followed by a discussion of the role of intervention itself in assisting international scientists. The principles that underlie the range of interventions are outlined briefly. It is also evident from the research that there are many stakeholders who should play a role in facilitating access to English-medium publishing. The role of the stakeholders and the rationale for their participation is outlined. Finally, there is a discussion of the impact on science when international participation is constrained or it is supported. In effect, we take a practical approach here to the implementation of policies and practical support for international scientists—all for the benefit of science itself.

9.2 THE CHALLENGES OF SECOND LANGUAGE SCIENCE WRITING IN ENGLISH

The findings of this study clearly indicate that scientists perceive it to be more difficult to write scientific articles in English than in their first language. In fact, this difficulty is measurable. The scientists reported that it is 24% more difficult to write a scientific article in English than to write a scientific article in their first language. There are other obstacles as well. Writing scientific articles in English produces more anxiety, and the final published article is less satisfying than writing in their first language. Their anxiety increases 22% when writing for publication in English, and their ultimate satisfaction decreases 11%. This state of affairs indicates that scientists must overcome substantial difficulties to participate in the scientific conversations of their discipline.

These findings oblige us to recognize the additional investment international scientists must make to become scientists. Anglophone scientists must learn their discipline through reading and writing activities that begin at the undergraduate level and continue through post-graduate education. The challenges of navigating this trajectory for Anglophone students in center countries are often discussed in higher education and science education studies. However, unlike Anglophone scientists, international scientists in peripheral countries have had to negotiate this trajectory in a foreign language: English. Thus, international scientists have an additional investment in their studies (Ammon, 2001). First, they have to learn the disciplinary content of their field, and additionally, they have to learn English. This doubling of the learning activities is a considerable burden, as was reported by many of the scientists in this study. The doubling of responsibilities required extra time devoted to English-language study, and greater cognitive demands to make sense of the scientific content when their English-language skills were still elementary. Secondly, international scientists must learn to produce articles, grant applications, and conference presentations in the foreign language. Again, their investment in these activities is greater than that of Anglophones.

Our findings support Ulrich Ammon's (2001) assertion that,

> It seems almost self-evident that the native speakers of the prevalent scientific language have less difficulty using it . . . and, therefore, have advantages over [non-native speakers] . . . It is easier for them to produce utterances and texts in line with the existing native speaker norms. (p. vii)

While Anglophones must learn to write scientific articles as part of their educational trajectory, they benefit from their inherent knowledge of "native speaker norms." Second language speakers of English may lack this inherent knowledge. For the scientists in this study, 20% reported that English continued to be a very significant barrier to their efforts to publish their scientific work. In summary, we can see that second language English-speaking scientists have a greater investment than native English-speaking scientists. First, they have had to learn a second language; second, they bear the additional costs of producing "linguistically adequate texts" (Ammon, 2001, p. 8).

This study demonstrates that producing linguistically adequate texts is not difficult for all non-native English-speaking scientists. In fact, an important finding of this study is that referring to second language speaking scientists as a single group is not accurate. In our study, we demonstrate that scientists at two neighboring institutions of higher education are different from one another. One institution is a public research university offering bachelor programs through doctoral degrees in the natural and applied sciences. The other institution is a research institute that offers only graduate degrees and where the science faculty have minimal teaching responsibilities.

The differences between the scientists in the two institutions are statistically significant across three principal factors. Those factors are their self-perceived difficulties in writing scientific articles in English, the anxiety associated with writing scientific articles in English, and their level of satisfaction with the final, published article in English.

Across these three factors, the scientists at the research institute reported fewer problems with writing scientific articles in English. They were more likely to rely on their abilities to write in English and rarely sought out translation services. Among this group of scientists, significantly, 29% reported that English presented no barrier to their ability to publish their work. The scientists based at the research institution perceive themselves to be more able to negotiate the publishing of their articles in English in comparison with the scientists based at the public university.

The overall experience of the scientists at the public university is less advantaged than those at the research institute. The scientists at the university reported greater challenges across the three factors of difficulty, anxiety, and satisfaction. Where the mean of the research institute scientists reported that the difficulty of writing research articles in English was 17% more difficult than writing research articles in Spanish, the scientists at the public university found writing in English to be 37% more difficult. Anxiety for the research institute scientists was rated at 16% greater when writing in English rather than Spanish, and anxiety increased to 31% for the teaching university faculty. The research institution scientists reported that their level of dissatisfaction with the published article in English was only 4% greater than in Spanish, but the teaching faculty were 21% more dissatisfied.

It is important to note that the scientists' reported levels of anxiety and difficulty differ in relation to producing scientific articles in English specifically. Writing articles in Spanish also involves a level of anxiety and difficulty, but the levels were not statistically different for the scientists between the two institutions. On seven-point scales, the level of difficulty regarding Spanish was 2.25 for university faculty and 2.57 for the research institute faculty; the level of anxiety was 2.56 for the university faculty and 2.54 for the research institute faculty. These figures indicate that writing research articles in Spanish does involve some difficulty and anxiety, but the levels are experienced as the same for all scientists. It is specifically writing in English that differs in the degree of challenge presented to the university faculty.

In addition to distinguishing faculty based on the kind of institution of higher learning in which they work, our data indicates that scientists with fewer years of experience (i.e., one to five years) differ from those scientists with more experience (i.e., six or more years). The scientists with fewer years of experience were termed "junior scientists," and those with more years of experience were termed "senior scientists." Our data showed that regarding publishing research articles in English, junior scientists experienced a significantly higher level of anxiety than senior scientists. This high level of anxiety is particularly important from a policy perspective. These junior scientists are the people who will become the established scientists of the future.

Based on the research reported in this book, it is evident that many scientists who are second language speakers of English face difficulties in publishing their scientific work in English. These scientists cannot be considered a homogenous group. Some scientists report that they require no assistance in their English publishing efforts. In our study, 15% of the 150 scientists fall into this category. All the others report that they do require some kind of assistance. For that reason, in Chapter 8, we presented a range of possible interventions that could be implemented. Those interventions are informed by a critical pragmatic approach, as outlined in the next section.

9.3 THE ROLE OF INTERVENTION

Supporting non-native English-speaking scientists in their writing of scientific articles has been a topic of some controversy among applied linguists. This controversy can be characterized as whether one supports a critical, a pragmatist, or a critical pragmatist approach.

Some applied linguists are highly critical of the expectation that international scientists should write like native speakers of English (Benesch, 2001; Canagarajah, 1996, 2002a). They exemplify the critical approach to scholarly writing. They argue that international contributions should be welcomed by journal editors and that the idiosyncrasies that mark an author as a second language speaker of English should not be altered or disparaged. The argument is one of inclusiveness of the forms of multilingual writing produced by researchers who are writing English as a second language. These arguments are based on critical appraisal of the hegemony of English-language genres and their conventions. From a critical perspective, these should be questioned and challenged.

In contrast, other applied linguists have emphasized the importance of writing like a native speaker. This is the "pragmatic" approach (Flowerdew, 2007). Linguistic support for the effacement of first language elements in a scientific manuscript is inherent in much of the

"English for specific purposes" and "English for academic purposes" literature (Allison, 1996: Swales, 1990). In this literature, international scholars are encouraged to analyze the texts of native speakers and use them as models for their writing. Applied linguists such as John Swales (1990; Swales & Feak, 2004) and Ken Hyland (2000a, Hyland & Bondi, 2006) have published detailed papers and monographs that report on the linguistic characteristics of scientific texts. They use varied methodologies, including corpus and genre analyses. In taking a pragmatic approach to non-Anglophone publishing, they imply that international scientists are more likely to have their manuscripts published when the linguistic and rhetorical construction of the manuscript is not questioned.

The third approach is the "critical pragmatic" approach, a term coined by Pennycook (1994, p. 317) and discussed by Flowerdew (2007). This approach stresses the importance of demanding "more equal opportunity for [non-Anglophone] writers" (Flowerdew, 2007, p. 22), and at the same time, encouraging training for such writers. Specifically when working with international writers, it is important that, first, they are made aware of the geopolitical, publishing-bibliometric and socio-cultural contexts that affect their writing. Second, they should be made aware of network possibilities and impediments that can affect their success in publishing (Lillis & Curry, 2010). Finally, they should be aware of the variable role that language itself can play in journal editors' decisions. There is evidence that language alone will not cause editors to reject a manuscript (Guardiano, Favilla, & Calaresu, 2007; Englander & López-Bonilla, 2011), whereas there are also documented cases where language shortcomings are the most salient features and lead to rejection (Englander & López-Bonilla, 2011). The critical pragmatic English for Academic Purposes (EAP) approach attempts to "synthesise the preoccupation with difference inherent in critical pedagogy and the preoccupation with access inherent in pragmatic pedagogy" (Harwood & Hadley, 2004, p. 366). It encourages international scholars to have "a critical mind set" (Flowerdew, 2007, p. 23), and at the same time, alert them "to the possible repercussions of some of the critical actions" (Flowerdew, 2007, p. 23) such as publishing in local languages, using language borrowing strategies, and submitting manuscripts with "non-standard English."

It is the critical pragmatic approach that we adopt in this book. The pragmatic approach is seen in several ways. We have documented through quantitative data the extra burden that most non-Anglophone scholars carry in publishing their work in English. We have presented qualitative data to demonstrate that scientists seek different kinds of support and assistance in order to better participate in English-language publishing practices. We have discussed the limitations that can accrue to scientists who do not participate in publishing in international English-language journals. We have reviewed research that situates issues of genre, writing, and publishing of scientific articles in English. These crucial ideas underpin the interventions that different stakeholders can provide to support international scientific scholarship. These interventions are based in a principled approach to supporting scientific writing in English.

9.4 Principles of Support

The interventions and policy recommendations that follow are based on a foundation of principles. A principled approach is crucial because it provides an empirically informed basis that policy makers can use when making decisions on how they want to implement programs of support for scientists who need to publish scientific articles in English. Seven principles serve as parameters for decision making, and they were presented in detail in Chapter 8. We list those principles here.

- *Long term commitment to writing education.* Facilitating the process of writing research articles in English as a second language needs to span the full length of higher education and professional work.
- *Differential needs and diversified educational interventions.* A range of approaches is necessary to respond to the developmental knowledge and different backgrounds of scientists.
- *Multilayered understanding of the research article.* The research article is one part of scientific disciplinary knowledge creation, and it occurs within a complex web of networks and publishing factors. It is not sufficient to only address linguistic factors in supporting English as a second language scientific writing.
- *Provision of expert support for science and writing.* Learning to write a research article involves developing skills and knowledge in relation to science and the literacy of science production. Ex-

pertise in both science and specialized science writing needs to be available.
- *Personalized, continual, and immediate support for research article writing.* Individualized support is needed for scientists in a framework that allows them immediate and continual access when and where they need it.
- *Demystification of the structures and processes of scientific publication.* Scientists need to be taught explicitly about the research article and the psychological and sociological processes involved.
- *Broad administrative, institutional, and financial support.* Administrators and institutions need to commit time and resources to the construction of an extended program that supports scientists in writing their research for English-medium publications when English is not their first language.

9.5 INTERVENTIONS BY DIFFERENT STAKEHOLDERS

The extensive interviews that we performed with scientists who speak English as a foreign language have provided a great deal of data concerning how they learned to write articles in English and the kinds of assistance that they seek. That data has been presented in detail in Chapters 6, 7, and 8. The results indicate that there are different kinds of desirable interventions, and those interventions can be implemented by different stakeholders involved in science education and professional scientific research. Those stakeholders are: (1) universities that educate scientists outside of the Anglophone countries; (2) makers of national, higher education policy; (3) the institutions where scientists perform their work; and (4) international, professional scientific bodies and associations. Each of these stakeholders is discussed below.

9.5.1 Universities

Universities that educate scientists outside of Anglophone countries have a crucial role to play in the eventual publishing success of their graduates. The results of our study, presented in Chapter 7, demonstrate a clear trajectory of learning events from the bachelor's degree through doctoral study (Figure 7.1). While we report specifically on the results obtained from our work in Mexico, we believe that they are relevant if contextualized to other parts of the non-Anglophone world. Our findings indicate that the trajectory of higher education represents common events at each stage in the scientist's education. Some learning events are conducted in the national language of Spanish, and others are conducted in English—a foreign language. At the earliest stages of the bachelor's degree, reading scientific texts in English is both important and difficult. Many of the scientists reported that reading scientific articles in English gave them access to the most accurate and up-to-date scientific content. However, there was virtually no support within the university to help the students learn to read these difficult texts in a foreign language. Most of the scientists had little knowledge of English at the time they entered the university. Additionally, the language courses that were usually available in the university were non-credit general English conversation classes. These classes were not directed at the needs of undergraduate students. Consequently, it seems obvious that universities can facilitate the eventual success of their students by integrating bilingual reading activities into the disciplinary curriculum. Integrating reading support for the genre and content knowledge of the discipline in both the local language and in English is desirable. Reading is critically important for the development of good writing skills (Bazerman, 2007).

Writing does not seem to get sufficient attention in the academic trajectory according to the scientists in this study. When writing does occur, the student-scientists report that there is often focused attention given to spelling and syntax without discussion of the larger organizational and rhetorical issues that are important. Insufficient attention seems to be directed at discussions of audience, argument structure, and strength of claim. Without guidance from expert scientists and faculty, the student-scientists are compelled to teach themselves autodidactically. This self-learning can be more frustrating and slower than explicit conversations with experienced writers. Therefore, greater emphasis on writing within educational disciplinary development is crucial.

The bilingual experience of peripheral student-scientists can be further developed as well. Virtually all of the writing produced at the undergraduate level is performed in the native language. In the transition to graduate studies, explicit transference to the genre demands of English would be helpful. Explorations of genre conventions in the native language and in English would allow student-scientists to increase their knowledge of the expectations regarding scientific writing in the two languages and also in their disciplinary cultures. This is especially important because some genres are likely to be written in the two languages once the scientist enters professional life. For example, conference abstracts, conference presentation slides, posters, grant applications, scientific articles, and correspondence with journal editors may be conducted in English or the native language, depending on the context and event.

Overall, universities can greatly facilitate the eventual publishing success of their graduates by broadening and deepening the reading and writing activities within course curricula. A number of American and Canadian universities have established such "writing across the disciplines" programs (see, for example, Poe, Lerner, & Craig, 2010; "WAC Clearinghouse," n.d.). These programs can serve as preliminary models for institutions of higher education around the world. However, international experience must still innovate better integration of "reading across the disciplines" and bilingual writing to fully prepare their graduates for professional scientific careers.

9.5.2 National Higher Education Policy Makers

Departments of higher education have the potential to support scientists in their efforts to publish their work in English. This can be done through incentives for publishing in English, monetary support for translation and editing services, and by providing opportunities for hosting and travelling to international conferences.

A number of countries specifically provide incentives for scientists who publish in English-language international journals. For example, Mexico and Turkey (Englander & Uzuner-Smith, forthcoming), Argentina (Ana Ines Heras, personal communication, June 30, 2009), Venezuela (Francoise Salager-Meyer, personal communication, January 18, 2010), and a number of European countries (Lillis & Curry, 2010; Moreno, 2010) have point systems for academic and scientific activities. Points are periodically calculated, and they qualify the scientist for monetary bonuses. Publishing in English-language journals are awarded a greater number of points than publishing in national or regional journals. This system of favoring English language rather than local publishing efforts has been criticized (Uzuner-Smith & Englander, 2011; Kent, 2009; Salager-Meyer, 2008); however, the incentive to publish in English to increase the visibility of scientific knowledge creation is valuable.

The difficulty of producing a scientific manuscript in English can be lessened when scientists can take advantage of professional translation and editing services. However, these services are often costly. There are companies based in the U.S. and Europe that specialize in these services. There may also be similar professional or informal services in many other countries around the world. Allocating funds through national departments of higher education, or through national science foundation agencies, can permit scientists to take advantage of these services. Translation and editing should be recognized as a legitimate expense for scientists around the world who undertake research.

Attending conferences allows access to international networks of scholars. Establishing networks has been shown to be crucial for peripheral scholars to participate in international publishing and international collaborative research (Lillis & Curry, 2010). The informal and formal conversations that occur at such venues open possibilities for peripheral scholars to get their work known. In these venues, introductions to established scientists based in center, Anglophone countries can facilitate the peripheral scholar's access to "insider" knowledge that facilitates effective participation in English-language publishing. Therefore, providing funds that allow scientists to travel to conferences is necessary. Similarly, providing funds that allow scientists to bring internationally recognized scientists to national conferences can provide crucial opportunities for conversations with established scientists. These conversations can help scientists navigate within the international publishing world. In short, national educational policy makers can create incentives and possibilities that support the scientists within their country to make international English-language publishing more easily obtainable.

9.5.3 Scientific and University Institutions

The institutions where scientists work can support professional scientists in many ways with their efforts to publish in English-medium international journals. In our view, institutions should consider such support to be an obligation for several reasons. First, we believe it is both unrealistic and unjust to expect every international scientist to develop "the secret language of academic work" to the level required to write publishable manuscripts in English as a second language (Pennycook, 1999, p. 330). International scientists must conduct credible research and then, additionally, produce sophisticated manuscripts in a foreign language. As demonstrated throughout this book, for the vast majority of international scientists, writing such manuscripts is a distinct burden. A scientist's participation in the disciplinary conversations of the field should not be so severely dependent on the individual scientist's writing ability in a foreign language. Secondly, institutions benefit from the scientific productivity of their faculty. Statistics that count the number of researchers in a given institution and the number of publications (and patents) produced by those researchers are increasingly common around the world (see publications of the OECD and UNESCO, for example). The statistics are used to rank institutions relative to one another, and prestige accrues to those institutions ranked highly (Uzuner-Smith & Englander, 2011). A consequence of the prestige is the institution's ability to attract the best and brightest students and recruit the best scientists for their faculty. The interrelationship of high international publishing and the best students and faculty serves to advance the institution's reputation. Finally, institutions may benefit economically from the scientific productivity of their faculty. These same statistical measures are sometimes counted in funding formulas from the federal government to individual institutions (Guadalupe Ortega Villa, personal communication, January 10, 2011). Thus, the productivity of the science faculty can directly impact the monetary resources of the institution.

In Chapter 8 of this volume, four interventions that can be implemented at the institutional level were outlined. Those interventions are: explicit teaching, collaboration in face-to-face interaction, expert and peer collaborations, and translation and editing services. Each of these interventions provides targeted support to scientists who seek assistance.

Very briefly, these interventions are:

- Explicit teaching is a structured set of courses or workshops that address topics relevant to writing and publishing scientific articles in English. Such courses can be delivered in a variety of formats, from face-to-face to online; they can vary in configuration, duration, type of evaluation, recognition awarded, and participants' investment. See Section 8.3.1 of Chapter 8 for a full discussion.

- Collaborative, face-to-face interaction with associates is a model of individual assistance for the scientist-author. Two types of associates are available to the author: (1) one who is an expert in providing rhetorical and linguistic support, and (2) one who is an expert in the scientific discipline. The scientist-author may seek out both types of associates, depending on the immediate project, for an ongoing series of meetings. These meetings occur in a specially designated location with appropriate resources. See Section 8.3.2 for further details.

- Expert and peer collaborations. Two program models that match experienced, published scientists with novice scientists are described in Section 8.3.3. One is a mentoring model in which newcomer scientists are matched with a senior colleague for an "insider" perspective on participating in the discipline. The other is the development of a faculty writing circle. In this model, following training in peer reviewing, interdisciplinary groups of faculty are formed. They meet on an ongoing basis to read, critique, and comment upon one another's manuscripts prior to journal submission. In both these models, the existing expertise of any one scientist is broadened through collaboration with peers and experts.

- Translation and editing services. This program of intervention establishes a series of services to assist scientist-authors. A translation service takes a manuscript produced in one language and reproduces it in another language. An editing service takes a manuscript produced in English by someone who is not a native speaker of English and revises it for linguistic and rhetorical acceptability. Both services function as a drop-off center for the author and the manuscript. See Section 8.3.4 for a more thorough discussion.

The institutions where scientists work can determine from the range of interventions presented what is feasible and desirable for their context. Faculty writing circles, for example, require little investment from the institution once the initial program is established. Similarly, various collaborations are possible with little direct monetary investment although release time and other forms of recognition may be required. A program of ongoing courses requires proper curriculum development and teaching expertise, and this can target general issues that pertain to scientific writing and publishing. Establishing a more permanent center that is available for writing support, translation, and editing is strongly recommended to alleviate the burden placed on individual scientists.

9.5.4 Professional Scientific Bodies and Associations

Calls for scientific bodies and associations to support international scientists' publishing efforts have been growing (Salager-Meyer, 2008). These bodies that host conferences and facilitate knowledge exchange can take an active role in welcoming and supporting multilingual scientists. This can be done by providing mentoring: matching senior or emeritus scientists with younger, international scientists. Such mentoring provides the international scientist with an insider view of the field. Additionally, it can open up networks that can lead to the collaborations and opportunities that facilitate greater participation. In many cases, scientific bodies and associations are the publishers of disciplinary journals, conference proceedings, and so on. Providing editorial assistance specifically to international scientists would facilitate a greater range of voices within the journal's pages.

In many cases, international representation on conference and editorial boards is dramatically small. For example, there were only two people from low-income countries among the 111 editorial board members of the five most important medical journals (Salger-Meyer, 2008). Expanding participation is important for broadening the range of perspectives and can lead to greater tolerance of idiosyncratic English writing and greater credibility accorded to institutions and their researchers outside of Anglophone countries.

In summary, there are institutions, bodies, and policy makers that have a substantial stake in the publishing success of international scientists. Each plays a crucial role in better assuring that the scientific work of scientists who speak English as a second or foreign language is published in English-medium journals.

9.6 POLICY IMPLICATIONS FOR SCIENCE

Scientific developments occur through conversations that build on existing knowledge. Those conversations are furthered through publication of scientific questions, findings, and discussions in peer-reviewed journals. To publish is to participate in the scientific conversations of one's field. Now, dissemination of that knowledge is tremendously dependent on English-language journals and English-language databases. Without adequate support, the work conducted by non-native English-speaking scientists, particularly those working outside the Anglophone countries, does not contribute to international conversations. The work "is lost to science" (Kaplan, 2001, p. 18).

International participation is necessary for the best furtherance of scientific thought. Scientists outside of Anglophone countries sometimes ask different questions, take different approaches, and take advantage of bibliographic data not available in English. The diversity of thought and approach is reflective of human diversity. Therefore, fundamentally, supporting international scientists in their efforts to write and publish in English as a second language as well as in their first language is a matter of fairness, justice, and equity. Such support furthers worldwide scientific knowledge creation.

Notes

Chapter 1

1. It should be noted that the authors of this book are not naïve in relation to the historical trajectory that has constructed the situation such that English has become the predominant language of science. We recognize the injustice of this situation, both in terms of linguistic access to English and the marginalization of non-English publication. However, we also recognize that this is the current situation and that access to participation needs to be addressed for the advancement of science and L2 scientists in particular.

Chapter 2

2. Junior researchers were defined as those people who had completed their Ph.D. within the previous five years and senior researchers were defined as having completed their Ph.D. six or more years prior to the time of the interviews.

3. All names are pseudonyms.

Chapter 7

4. Dr. Sanchez is a senior researcher at the research institute.
5. Drs. Rodriguez and Lorenzo are both senior researchers at the university.
6. Dr. Juarez is a senior researcher at the research institute.
7. Dr. Jornado is a junior researcher at the university.
8. Dr. Carrillo is a junior researcher at the research institute.
9. Dr. Gonzalez is a junior researcher at the university.
10. Dr. Zorrillo is a senior researcher at the university.
11. Dr. Ramos is a junior researcher at the university.
12. Dr. Salinas is a junior researcher at the research institute.
13. Dr. Gomez is a senior researcher at the university.
14. Dr. Ortiz is a senior researcher at the research institute.
15. Dr. Papel is junior researcher at the research institute.
16. Dr. Alamo is a senior researcher at the research institute.
17. Dr. Ilana is a junior researcher at the research institute.

References

Abdollahzadeh, E. (2011). Poring over the findings: Interpersonal authorial engagement in applied linguistics papers. *Journal of Pragmatics, 43*(1), 288–297.

Ahmad, U.K. (1997). Research article introductions in Malay: Rhetoric in an emerging research community. In A. Duszak (Ed.), *Culture and styles in academic discourse* (pp. 273–303). Berlin, Germany: Mouton de Gruyter.

Allison, D. (1996). Pragmatist discourse and English for Academic Purposes. *English for Specific Purposes, 15*(2), 85–103.

Ammon, U. (1998). *Ist Deutsch noch internationale Wissenschaftssprache? Englisch auch fir die Lehre an den deutschsprachigen Hochschulen.* Berlin: Mouton de Gruyter.

Ammon, U. (2001). *The dominance of English as a language of science: Effects on other languages and language communities.* Berlin: Mouton de Gruyter.

Ammon, U. (2006). Language planning for international scientific communication: An overview of questions and potential solutions. *Current Issues in Language Planning, 7* (1), 1–30.

Aydinli, E., & Mathews, J., (2000). Are the core and periphery irreconcilable? *International Studies Perspectives, 1*(3), 289–303.

Bazerman, C. (1988). *Shaping written knowledge: The genre and activity of the experiential article in science.* Madison: University of Wisconsin Press.

Bazermen, C. (Ed.). (2007). *Handbook of research on writing.* New York: Routledge.

Belcher, D. (1994). The apprenticeship approach to advanced academic literacy: Graduate students and their mentors. *English for Specific Purposes, 13*(1), 23–34.

Belcher, D. (2007). Seeking acceptance in an English-only research world. *Journal of Second Language Writing, 16*(1), 1–22.

Belcher, W.L. (2009). *Writing your journal article in 12 weeks: A guide to academic publishing success.* Thousand Oaks, CA: Sage.

Belcher, D., & Conner, U. (2001). *Reflections on multiliterate lives.* Clevedon: Multilingual Matters.

Benesch, S. (2001). *Critical English for academic purposes: Theory, politics and practice.* Mahwah, NJ: Erlbaum.

Berkenkotter, C., & Huckin, T. (1995). *Genre knowledge in disciplinary communication: Cognition, culture, power.* Mahwah, NJ: Erlbaum.

Bhatia, V.K. (2004). *Worlds of written discourse: A genre-based view.* London: Continuum.

Biber, D., Conrad, S., & Cortes, V. (2004). If you look at . . . : Lexical bundles in university teaching and textbooks. *Applied Linguistics, 25*(3), 371–405.

Bordons, M., Fernández, M.T., & Gómez, I. (2002). Advantages and limitation in the use of impact factor measures for the assessment of research performance in a peripheral country. *Scientometrics, 53*(2), 195–206.

Burgess, S. (2002). Packed houses and intimate gatherings: Audience and rhetorical structure. In J. Flowerdew (Ed.), *Academic discourse* (pp. 196–215). Harlow: Longman.

Burrough-Boenisch, J. (2003). Shapers of published NNS research articles. *Journal of Second Language Writing, 12*(3), 223–243.

Byrd, P., & Coxhead, A. (2010). On the other hand: Lexical bundles in academic writing and in the teaching of EAP. *University of Sydney Papers in TESOL, 5*, 31–64.

Byrnes, H. (2002). Toward academic-level foreign language abilities: Reconsidering foundational assumptions concerning pedagogical options. In B.L. Leaver, & B. Shekhtman (Eds.), *Developing professional-level language proficiency* (pp. 34–58). Cambridge, UK: Cambridge University Press.

Canagarajah, A.S. (1996). "Nondiscursive" requirements in academic publishing, material resources of periphery scholars, and the politics of knowledge production." *Written Communication, 13*(4), 435–472.

Canagarajah, A.S. (2002a). *A geopolitics of academic writing.* Pittsburgh: University of Pittsburgh Press.

Canagarajah, A.S. (2002b). *Critical academic writing and multilingual students.* Ann Arbor, MI: University of Michigan Press.

Cargill, M. (2011). *Collaborative interdisciplinary publication skills education: Implementation and implications in international science research contexts.* (Unpublished doctoral dissertation). University of Adelaide, Australia.

Cargill, M., & O'Connor, P. (2009). *Writing scientific research articles: Strategy and steps.* Oxford: Wiley-Blackwell.

Cargill, M., O'Connor, P., & Li, Y. (2012). Educating Chinese scientists to write for international journals. *English for Specific Purposes, 31*(1), 60–69.

Centro de Información y Documentación Científica (CINDOC). (1999). El español en las revistas de ciencias y tecnología recogidas en ocho bases de datos internacionales. Retrieved from http://cvc.cervantes.es/obref/anuario/anuario_99.

Cho, S. (2004). Challenges of entering discourse communities through publishing in English: perspectives of nonnative-speaking doctoral students

in the United States of America. *Journal of Language, Identity and Education, 3*(1), 47–72.
Cho, D.W. (2009). Science journal paper writing in an EFL context: The case of Korea. *English for Specific Purposes, 28*(4), 230–239.
CONACYT. (2009). *Informe General del Estado de la Ciencia y la Technologia, 2008* [General report of the state of science and technology, 2008]. Mexico City: Mexico.
Cortes, V. (2010). Discourse on the move: Using corpus analysis to describe discourse structure. *English for Specific Purposes, 29*(1), 70–73.
Craig, J., Poe, M., & González-Rojas, M. (2010). Professional communication in a global context: A collaboration between the Massachusetts Institute of Technology, Instituto Tecnológico y de Estudios Superiores de Monterrey, Mexico, and Universidad de Quintana Roo, Mexico. *Journal of Business and Technical Communication, 24*(3), 267–295.
Curry, M.J. (2011). Book reviews: Publish or perish. *English for Academic Purposes, 10*, 64–67.
Curry, M.J., & Lillis, T. (2004). Multilingual scholars and the imperative to publish in English: Negotiating interests, demands, and rewards. *TESOL Quarterly, 38*(4), 663–688.
Curry, M.J., & Lillis, T. (2010). Academic research networks: Accessing resources for English-medium publishing. *English for Specific Purposes, 29*(4), 281–295.
Dong, Y.R. (1996). Learning how to use citations for knowledge transformation: Non-native doctoral students' dissertation writing in science. *Research in the Teaching of English, 30*(4), 428–457.
Duff, P. (2010). Language socialization into academic discourse communities. *ARAL, 30*, 169–192.
Duszak, A. (Ed.). (1997). *Culture and styles of academic discourse.* Berlin: Mouton de Gruyter.
Duszak, A., & Lewkowicz, J. (2008). Publishing academic texts in English: A Polish perspective. *English for Academic Purposes, 7*, 108–120.
El Malik, A.T., & Nesi, H. (2008). Publishing research in a second language: The case of Sudanese contributors to international medical journals. *English for Academic Purposes, 7*(2), 87–96.
Englander, K. (2006). Revision of scientific manuscripts by nonnative-English-speaking scientists in response to journal editors' language critiques. *Journal of Applied Linguistics, 3*(2), 129–161.
Englander, K. (2009). Transformation of the identities of nonnative English speaking scientists. *Journal of Language, Identity and Education 8*(1), 35–53.
Englander, K. (2010). But it would be good in Spanish: An analysis of awkward scholarly writing in English by L2 writers. In S. Santos (Ed.),

EFL writing in Mexican universities: Research and experience (pp.55–71). Nayarit: Universidad Autónoma de Nayarit.

Englander, K. & Uzuner-Smith, S. (Forthcoming). The role of policy in constructing the peripheral scientists in the era of gloabalization. *Language Policy*.

Englander, K., & López-Bonilla, G. (2011). Acknowledging or denying membership: Reviewers' responses to non-Anglophone scientists' manuscripts. *Discourse Studies, 13*(5), 1–23.

Ferguson, G. (2007). The global spread of English, scientific communication and ESP: Questions of equity, access and domain loss. *Ibérica, 13*, 7–38.

Fernández Polo, F.J. (1999). *Traducción y retórica contrastiva: A propósito de la traducción de textos de divulgación científica del inglés al español* [Translation and contrastive rhetoric: a proposal for the translation of popular texts from English to Spanish]. Santiago de Compostela, Spain: Universidad de Santiago de Compostela Servicio de Publicacións.

Flowerdew, J. (1999a). Writing for scholarly publication in English: The case of Hong Kong. *Journal of Second Language Writing, 8*(2), 123–145.

Flowerdew, J. (1999b). Problems in writing for scholarly publication in English: The case of Hong Kong. *Journal of Second Language Writing, 8*(3), 243–264.

Flowerdew, J. (2000). Discourse community, legitimate peripheral participation, and the nonnative-English-speaking scholar. *TESOL Quarterly, 34*(1), 127–150.

Flowerdew, J. (2001). Attitudes of journal editors to nonnative speaker contributions. *TESOL Quarterly, 35*(1), 121–150.

Flowerdew, J. (2007). The non-Anglophone scholar on the periphery of scholarly publication. *AILA Review, 20*(1), 14–27.

Flowerdew, J., & Li, Y. (2009). English or Chinese? The trade-off between local and international publication among Chinese academics in the humanities and social sciences. *Journal of Second Language Writing, 18*(1), 1–16.

Flowerdew, J., & Li, Y. (2007). Language re-use among Chinese apprentice scientists writing for publication. *Applied Linguistics, 28*(3), 440–463.

Fortanet, I. (2008). Evaluative language in peer review referee reports. *English for Academic Purposes, 7*(1), 27–37.

Fox, H. (1994). *Listening to the world: Cultural issues in academic writing*. Urbana, IL: NCTE.

Franck, G. (1998). *Ökonomie der Aufmerksamkeit, ein Entwurf*. Munchen: Carl Hanser Verlag.

Franck, G. (1999). Scientific communication—vanity fair? *Science, 286*(5437), 53–55.

Galison, P., & Helvy, B. (1992). *Big science: The growth of large-scale research*. Stanford, CA: Stanford University Press.

Garfield, E. (1965). Can citation indexing be automated? In M.E. Stevens, V.E. Giuliano, & L.B. Heilprin (Eds.), *Statistical association methods for mechanized documentation* (pp. 186–194). Washington, D.C.: National Bureau of Standards.

Geertz, C. (1973). *The interpretation of cultures.* New York: Basic Books.

Gibbs, W. (1995). Trends in scientific communication: Lost science in the third world. *Scientific American*, August, 76–83.

Glasman-Deal, H. (2010). *Science research writing for non-native speakers of English.* London: Imperial College Press.

González-Reyna, M.F. (2010). *Teachers' role in forming students' literacy practices—in Spanish and English—in an undergraduate scientific discipline in a Mexican public university.* (Unpublished master's thesis). Universidad Autónoma de Baja California, Mexico.

Gordon, M.D. (1980). The role of referees in scientific communication. In J. Hartley (Ed.), *The psychology of written communication: Selected readings* (pp. 263–275). London: Kogan Page.

Gosden, H. (1992). Research writing and NNSs: From the editors. *Journal of Second Language Writing, 1*(2), 123–139.

Gosden, H. (1996). Verbal reports of Japanese novices' research writing practices in English. *Journal of Second Language Writing, 5*(2), 109–128.

Gosden, H. (2003). "Why not give us the full story?": Functions of referees' comments in peer review of scientific research papers. *English for Academic Purposes, 2*, 87–101.

Gross, A.G. (1990). *The rhetoric of science.* (2nd ed.). Cambridge: Harvard University Press.

Gross, A. (2006). *Starring the text: The place of rhetoric in science studies.* Carbondale: Southern Illinios University Press.

Gross, A.G., Harmon, J.E., & Reidy, M.S. (2009). *Communicating science: The scientific article from the 17th century to the present.* West Lafayette: Parlor Press.

Guardiano, C., Favilla, M.E., & Calaresu, E. (2007). Stereotypes about English as the language of science. *AILA Review, 20*, 28–52.

Hamel, R.E. (2006). Spanish in science and higher education: Perspectives for a plurilingual language policy in the Spanish speaking world. *Current Issues in Language Planning , 7*(1), 95–125.

Hamel, R.E. (2007). The dominance of English in the international scientific periodical literature and the future of language use in science. *AILA Review, 20*, 53–71.

Hanauer, D.I. (1998). The genre-specific hypothesis of reading: Reading poetry and reading encyclopedic items. *Poetics, 26*(2), 63–80.

Hanauer, D.I. (2004). Silence, voice and erasure: Psychological embodiment in graffiti at the site of Prime Minister Rabin's assassination. *Psychotherapy in the Arts, 31*(1), 29–35.

Hanauer, D.I. (2006). *Scientific discourse: Multiliteracy in the classroom*. London: Continuum Press.

Hanauer, D.I., & Englander, K. (2011). Quantifying the burden of writing research articles in a second language: Data from Mexican scientists. *Written Communication, 28*(4), 403–416.

Hanauer, D.I., Hatfull, G.F., & Jacobs-Sera, D. (2009). *Active assessment: Assessing scientific inquiry*. New York, NY: Springer.

Hansen, K. (1998). *A rhetoric for the social sciences: A guide for academic and professional communication*. Upper Saddle River: Prentice Hall.

Harris, R.A. (Ed.). (1997). *Landmark essays on rhetoric of science: Case studies*. Mahwah, NJ: Erlbaum.

Hartley, J. (2008). *Academic writing and publishing: A practical handbook*. London: Routledge.

Harwood, N. & Hadley, G. (2004). Demystifying institutional practices: Critical pragmatism and the teaching of academic writing. *English for Specific Purposes, 23*, 355–377.

Hildebrand, G. (1998). Disrupting hegemonic writing practices in school science: Contesting the right way to write. *Journal of Research in Science Teaching, 35*(4), 345–362.

Hinds, J. (1987). Reader vs. writer responsibility: A new typology. In U. Connor, & R.B. Kaplan (Eds.), *Writing across languages: Analysis of L2 texts* (pp. 141–152). Reading, MA: Addison-Wesley.

Hoffman, R. (1988). Under the surface of the chemical article. *Angewandte Chemie: International Edition in English, 27*, 1593–1602.

Hopkins, A., & Dudley-Evans, T. (1988). A genre-based investigation of the discussion sections in articles and dissertations. *English for Specific Purposes, 7*(2), 113–121.

Huang, J.C. (2010). Publishing and learning writing for publication in English: Perspectives of NNES PhD students in science. *English for Academic Purposes, 9*(1), 33–44.

Hwang, K. (2005). The inferior science and the dominant use of English in knowledge production: A case study of Korean science and technology. *Science Communication, 26*(4), 390–427.

Hwang, K. (2008). International collaboration in multilayered center-periphery in the globalization of science and technology. *Science, Technology & Human Values, 33*(1), 101–133.

Hyland, K. (2000a). *Disciplinary discourses: Social interactions in academic writing*. London: Longman.

Hyland, K. (2000b). Hedges, boosters and lexical invisibility: Noticing modifiers in academic texts. *Language Awareness. 9*(4), 179–197.

Hyland, K. (2005a). *Metadiscourse*. London: Continuum.

Hyland, K. (2005b). Stance and engagement: A model of interaction in academic discourse. *Discourse Studies, 7*(2), 173–191.

Hyland, K. (2008). As can be seen: Lexical bundles and disciplinary variation. *English for Specific Purposes, 27*(1), 4–21.
Hyland, K. (2009). *Academic discourse.* London: Continuum.
Hyland, K., & Bondi, M. (Eds.). (2006). *Academic discourse across disciplines.* Bern: Peter Lang.
Hyland, K., & Salager-Meyer, F. (2008). Scientific writing. *Annual Review of Information Science and Technology, 42*(1), 297–338.
Ivanic, R. (1998). *Writing and identity: The discoursal construction of identity in academic writing.* Amsterdam: Benjamins.
Jaffe, S. (2003). No pardon for poor English in science. *The Scientist,* (March 3), 44–45.
Jaeggli, O. (1986). Passive. *Linguistic Inquiry, 17,* 582–622.
Kachru, B.B. (1996). The paradigms of marginality. *World Englishes, 15*(3), 241–255.
Kachru, B.B. (1997). World Englishes and English-using communities. *ARAL, 17,* 66–87.
Kamberelis, G. (1995). Genre as instituitional informed social practice. *Journal of Contemporary Legal Practice, 6,* 115–171.
Kaplan, R.B. (2001). English—The accidental language of science? In U. Ammon (Ed.), *The dominance of English as a language of science: Effects on other languages and language communities* (pp. 3–26). Berlin: Mouton de Gruyter.
Katz, M.J. (2009). *From research to manuscript: A guide to scientific writing.* (2nd ed.). New York, NY: Springer.
Katz, J.S., & Martin, B. R. (1997). What is research collaboration? *Research Policy 26,* 1–18.
Kennedy, D. (1997). *Academic duty.* Cambridge, MA: Harvard University Press.
Kent, R. (2009). *Políticas de educación superior en México durante la modernización* [Policies of higher education in Mexico during modernization]. Mexico City: ANUIES.
Kerans, M.E. (2001). Eliciting substantive revision of manuscripts for peer review through process-oriented conferences with Spanish scientists. In C. Muñoz (Ed.), *Trabajos en lingüística aplicada* [Essays in applied linguistics] (pp. 339–347). Barcelona: Universitat de Barcelona.
King, D.A. (2004). The scientific impact of nations. *Nature, 430* (July 15), 311–316.
Knorr-Cetina, K. (1981). *The manufacture of knowledge.* Oxford: Pergamum Press.
Kourilová, M. (1996). Interactive functions of language in peer reviews of medical papers written by non-native users of English. *Unesco ALSED-LSP newsletter, 19*(1), 4–21.

Kuhn, T. (1962). *The structure of scientific revolutions.* Chicago: University of Chicago Press

Kumaravadivelu, B. (2002). *Beyond methods: Macrostrategies for language teaching.* New Haven: Yale University Press.

Kuznetsov, Y., & Dahlman, C. (2008). *Mexico's transition to a knowledge-based economy.* Washington, D.C.: The World Bank.

La Madeleine, B.L. (2007, 25 Jan.). Lost in translation. *Nature, 454*–455.

Latour, B, & Woolgar, S. (1986). *Laboratory life: The construction of scientific fact.* Princeton: Princeton University Press.

Li, Y. (2002). Writing for international publication: The perception of Chinese doctoral researchers. *Asian Journal of English Language Teaching, 12*, 179–193.

Li, Y. (2005). Multidimensional enculturation: The case of an EFL Chinese doctoral student. *Journal of Asian Pacific Communication, 15*, 153–70.

Li, Y. (2006). A doctoral student of physics writing for publication: A socio-politically-oriented case study. *English for Specific Purposes, 25*(4), 456–478.

Li, Y., & Flowerdew, J. (2007). Shaping Chinese novice scientists' manuscripts for publication. *Journal of Second Language Writing, 16*(2), 100–117.

Lillis, T., & Curry, M.J. (2006a). Professional academic writing by multilingual scholars: Interactions with literacy brokers in the production of English-medium texts. *Written Communication, 23*(1), 3–35.

Lillis, T., & Curry, M.J. (2006b). Reframing notions of competence in scholarly writing: From individual to networked activity. *Revista Canaria de Estudios Ingleses, 53* (November), 63–78.

Lillis, T., & Curry, M.J. (2010). *Academic writing in a global context: The politics and practices of publishing in English.* London: Routledge.

Link, A.M. (1998). U.S. and Non-U.S. submissions: An analysis of reviewer bias. *JAMA, 280*(3), 246–247.

Liu, J. (2004). Co-constructing academic discourse from the periphery: Chinese applied linguists' centripetal participation in scholarly publication. *Asian Journal of English Language Teaching, 14*, 122–131.

Mair, C., & Hundt, M. (1995). Why is the progressive becoming more frequent in English? A corpus-based investigation of language change in progress. *Zeitschrift für Anglistik und Amerikanistik, 43*(2), 111–122.

Man, J.P., Weinkauf, J.G., Tsang, M., & Sin, D.D. (2004). Why do some countries publish more than others? An international comparison of research funding, English proficiency and publication output in highly ranked general medical journals. *European Journal of Epidemiology, 19*, 811–817.

Martel, A. (2001). When does knowledge have national language? Language policy-making for science and technology. In U. Ammon (Ed.), *The dom-*

inance of English as a language of science: Effects on other languages and language communities (pp. 27–58). Berlin: Mouton de Gruyter.

Martín, M.P. (2003). A genre analysis of English and Spanish research paper abstracts in experimental social sciences. *English for Specific Purposes, 22*(1), 25–43.

Martín-Martín, P., & Burgess, S. (2004). The rhetorical management of academic criticism in research article abstracts. *Text, 24*(2), 171–195.

Martínez, I.A. (2001). Impersonality in the research article as revealed by analysis of the transitivity structure. *English for Specific Purposes, 20*(3), 227–247.

Martínez, I.A. (2003). Aspects of theme in the method and discussion sections of biology journal articles in English. *Journal of English for Academic Purposes, 2*(2), 103–123.

Matsuda, P.K. (2008). *Second Language Writing*. Invited lecture, Indiana University of Pennsylvania.

Matsumoto, K. (1995). Research paper writing strategies of professional Japanese EFL writers. *TESL Canada Journal, 13*(1), 17–27.

McGinty, S. (1999). *Gatekeepers of knowledge: Journal editors in the sciences and the social sciences*. Westport, CN: Bergin & Garvey.

McGinn, M. & Roth, W. (1999). Preparing students for competent scientific practice: Implications of recent research in science and technology studies. *Educational Researcher, 28*(3), 14–24.

McGrail, M.R., Rickard, C.M., & Jones, R. (2006). Publish or perish: A systematic review of interventions to increase academic publication rates. *Higher Education Research & Development, 25*(1), 19–35.

Mervis, J. (2010). Handful of U.S. schools claim larger share of output. *Science, 30*, 1032.

Miller, C.R. (1984). Genre as social action. *Quarterly Journal of Speech, 70*, 151–167.

Moll, T.M. (2009). A presentation course design for academics of English as an additional language: A multimodal approach. In S. Burgess & P. Martín-Martín (Eds.), *English as an additional language in research publication and communication*, (pp. 237–253). Bern, Switzerland: Peter Lang.

Moreno, A.I. (2010). Researching into English for research publication purposes from an applied intercultural perspective. In M.F. Ruiz-Garrido, J.C. Palmer-Silveira, & I. Fortanet-Gómez (Eds.), *English for Professional and Academic Purposes*, (pp. 57–71). Amsterdam: Rodopi

Mungra, P., & Webber, P. (2010). Peer review process in medical research publications: Language and content comments. *English for Specific Purposes, 29*(1), 43–53.

Murphy, C., & Sherwood, S. (2011). *The St. Martin's sourcebook for writing tutors* (4th ed.). Bedford/St. Martins.

Murray, H., & Dingwall, S. (2001). The dominance of English at European universities: Switzerland and Sweden compared. In U. Ammon (Ed.), *The dominance of English as a language of science: Effects on other languages and language communities* (pp. 85–112). Berlin: Mouton de Gruyter.

Murray, R. (2009). *Writing for academic journals* (2nd ed.). Berkshire, UK: McGraw Hill/Open University Press.

Myers, G. (1990). *Writing biology*. Madison: University of Wisconsin Press.

Nakamura, J. Shernoff, D., & Hooker, C. (2009). *Good mentoring: Fostering excellent practice in higher education*. San Francisco, CA: Jossey-Bass.

Nygaard, L. (2009). *Writing for scholars: A practical guide to making sense and being heard*. Oslo: Universitetsforlaget/Copenhagen Business School Press/Liber.

Oda, M. (2007). "Globalization or the world in English: Is Japan ready to face the waves?" *International Multilingual Research Journal, 1*(2), 119–126.

Olsen, L. A., & Huckin, T. N. (1991). *Technical writing and professional communication* (2nd ed.). New York: McGraw-Hill.

Organisation for Economic Cooperation and Development (OECD). (2009). *Higher Education to 2030, Volume 2, Globalisation*. OECD.

Pendlebury, D. A. (2008). *White paper: Using bibliometrics in evaluating research*. Retrieved from http://thomsonreuters.com/content/science/pdf/ssr/training/UsingBibliometricsinEval_WP.pdf

Pennycook, A. (1994). *The cultural politics of English as an international language*. London: Longman..

Pennycook, A. (1999). Introduction: Critical approaches to TESOL. *TESOL Quarterly, 33*(3), 329–348.

Penrose, A.M., & Katz, S.B. (2010). *Writing in the sciences: Exploring conventions of scientific discourse*. New York: Longman.

Pérez-Llantada, C. (2007). Native and non-native English scholars publishing research internationally: A small-scale study on authorial (in)visibility. *Journal of Applied Linguistics, 4*(2), 217–237.

Pérez-Llantada, C. (2010). The discourse functions of metadiscourse in published academic writing: Issues of culture and language. *Nordic Journal of English Studies, 9*(2), 41–68.

Pérez-Llantada, C., Plo, R., & Ferguson, G.R. (2011). "You don't say what you know, only what you can": The perceptions and practices of senior Spanish academics regarding research dissemination in English. *English for Specific Purposes, 30*(1), 18–30.

Poe, M, Lerner, N., & Craig, J. (2010). *Learning to communication in science and engineering: Case studies from MIT*. Cambridge, MA: MIT Press.

Preisler, B. (2005). Deconstructing "the domain of science" as a sociolinguistic entity in EFL societies: The relationship between English and Danish in higher education and research. In B. Preisler, A. Fabricius, H. Haberland, S. Kjaerbeck, & K. Risager (Eds.), *The consequences of mobility* (pp. 238–248). Roskilde: Roskilde University.

Prelli, L. (1989). *A rhetoric of science: Inventing scientific discourse*. Columbia: University of South Carolina Press.
Prior, P. (1998). *Writing disciplinarity: A sociohistoric account of literate activity in the academy*. Mahwah, NJ: Erlbaum.
Rankin, E. (2001). *The work of writing: Insights and strategies for academics and professionals*. San Francisco: Jossey-Bass.
Reid, N. (2010). *Getting published in international journals*. Oslo: NOVAd-Norwegian Social Research.
St. John, M.J. (1987). Writing processes of Spanish scientists publishing in English. *English for Specific Purposes, 6*(2), 113–120.
Salager-Meyer, F. (2008). Scientific publishing in developing countries: Challenges for the future. *Journal of English for Academic Purposes, 7*(2), 121–132.
Salager-Meyer, F., Defives, G., & Hamelinsck, M. (1996). Epistemic modality in 19th and 20th century medical English written discourse: A principal component analysis. *Interface. Journal of Applied Linguistics, 11*(2), 95–117.
Seoane, E. (2006). Changing styles: On the recent evolution of scientific British and American English. In C. Dalton-Puffer, D. Kastovsky, N. Ritt, & H. Schendly (Eds.), *Syntax, style and grammatical norms: English from 1500–2000* (pp. 191–211). Frankfurt: Peter Lang.
Shashok, K. (1992). Educating international authors. *European Science Editing, 45*, 5–7.
Silva, T. (1993). Toward an understanding of the distinct nature of L2 writing: The ESL research and its implications. *TESOL Quarterly, 27*(4), 657–675.
Sionis, C. (1995). Communication strategies in the writing of scientific research articles by non-native users of English. *English for Specific Purposes, 14*(2), 99–113.
Storch, N., & Tapper, J. (2009). The impact of an EAP course on postgraduate writing. *English for Academic Purposes, 8*(3), 207–223.
Stratton, C.R. (1984). *Technical Writing: Process and Product*. New York: Holt, Rinehart, and Winston.
Swales, J.M. (1990). *Genre analysis: English in academic and research settings*. Cambridge: Cambridge University Press.
Swales, J.M. (2002). On models in applied discourse analysis. In C.N.Candlin (Ed.), *Research and practice in professional discourse* (pp. 61–77). Kowloon, Hong Kong: City University of Hong Kong Press.
Swales, J. (2004). *Research Genres: Exploration and Applications*. Cambridge: Cambridge University Press.
Swales, J.M., & Feak, C. (2004). *Academic writing for graduate students: Essential tasks and skills* (2nd ed.). Ann Arbor, MI: University of Michigan Press.

Tardy, C.M. (2005). "It's like a story": Rhetorical language development in advanced academic literacy. *English for Academic Purposes, 4*(4), 325–338.

Tardy, C.M., & Matsuda, P.K. (2009). The construction of author voice by editorial board members. *Written Communication, 26*(1), 32–52.

Testa, J. (2012, May). The Thompson scientific journal selection process. Retrieved from http://scientific.thomson.com/free/essays/selectionofmaterial/journalselection/.

Tsunoda, M. (1983). Les langues internationales dans les publications scientifique. Le francais et les langue scientifiques de demain. *Sophia Linguistica, 13*, 144–155.

Uzuner, S. (2008). Multilingual scholars' participation in core/global academic communities: A literature review. *English for Academic Purposes, 7*(4), 250–263.

Uzuner-Smith, S., & Englander, K. (2011). *Constructing the peripheral scientist in the era of globalization.* Paper presented at American Association of Applied Linguistics, Chicago, IL.

Ventola, E. (1991). (Ed.). *Functional and systemic linguistics.* Berlin: Mouton de Gruyter.

Wagner, C.S., & Leydesdorff, L. (2003). Mapping the network of global science: Comparing international co-authorships from 1990 to 2000. Retrieved from http://users.fmg.uva.nl/lleydesdorff/cwagner/Thesis/Chapter%20V.Global%20mapping.pdf.

Wallerstein, I. (1991). *Geopolitics and geoculture.* Cambridge: Cambridge University Press.

Wanner, A. (2009). *Deconstructing the English passive.* Berlin: Walter de Gruyter.

Weinberg, A. (1967). *Reflections on big science.* Cambridge: MIT Press.

Wenneras, C., & Wold, A. (1997, 22 May). Nepotism and sexism in peer-review. *Nature, 387,* 341–343.

Williams, M. (1999). *Style: Ten lessons in clarity and grace* (4th ed.). New York: HarperCollins.

Winsor, D. (1996). *Writing like an engineer: A rhetorical education.* London: Routledge.

Yaffe, M.B. (2009). Re-reviewing peer review. *Science Signaling, 2*(85), 1–2.

Yakhontova, T. (2006). Cultural and disciplinary variation in academic discourse: The issue of influencing factors. *English for Academic Purposes, 5*(2), 153–167.

Ynalvez, M.A., & Shrum, W.M. (2009). International graduate science training and scientific collaboration. *International Sociology, 24*(6), 870–901.

Zerbe, M.J. (2007). *Composition and the rhetoric of science: Engaging the dominant discourse.* Carbondale: Southern Illinois University Press

Index

academic brokers, 21, 38
Academic Word, 49
academic writing, 11, 47, 50, 53, 88, 101, 117, 143, 145, 164
active voice, 24
adjective, 29
administrative service, 60, 108, 134, 139, 167
administrative support, 139
Ammon, Ulrich, 3, 6, 11, 49, 155, 161–162
American Journal Experts, 49
Anglophone countries, 33–34, 36–38, 40–42, 46–47, 51, 53, 160–161, 167, 170, 173–174
ANOVA, 27
anxiety, 61–62, 68–72, 76–82, 129–130, 133, 149, 161–164
applied linguistics, 14, 32, 40, 42–44, 140
assessment, 4
author, 4, 20, 28, 35, 41 43, 45, 52, 56, 85–86, 89–90, 94, 97, 100, 102–103, 106, 109, 133, 144–146, 149–151, 155–156, 164, 172; co–author, 84–85, 90, 94, 97, 102–104, 117–118, 120, 124; first author, 84, 88, 90, 94, 97, 100, 104; second author, 84, 86–90, 97
autodidactic, 115–116, 118–119, 124

bibliometric, 18–19, 34, 36, 56–57; data, 19, 34, 36; measures, 18
bilingual, 50, 85, 122, 145, 155, 168–169
Biochemistry, 45
Bioscience Writers, 49

Cargill, Margaret, 137–138, 142–143
Center for Scientific Research and Higher Education of Ensenada (CICESE), 59

citation, 17–19, 23
co–author, 84–85, 90, 94, 97, 102–104, 117–118, 120, 124
coherence, 52, 94, 127
collaboration, 36, 38–39, 49, 52–53, 59, 66, 90, 96–100, 102, 104, 106, 109, 119–120, 123, 143–147, 150, 154, 158–159, 170–172
Collaborative Interdisciplinary Publication Skills Education, 143
collaborative writing, 98, 100, 102, 123–124, 133
conferences, academic, 87, 91, 94, 100, 102, 106, 108, 117, 127, 144, 154, 162, 169, 173
Content Delivery Matrix, 141–142
content module, 141
corpus linguistics, 51, 53, 128
Create a Research Space (CARS), 26, 51
critical–pragmatic approach, 11–12
Curry, Mary Jane, 7, 21, 33, 35, 37–38, 40, 42, 47, 50, 119, 139, 143, 155, 165, 169–170

data, 6–8, 13, 17–20, 23–25, 27, 29–30, 32, 34–35, 40, 62–63, 66–67, 70, 74, 76, 80, 82–84, 86, 96, 109–112, 121, 127, 130, 134, 136, 138–140, 163, 167, 174; descriptive, 68, 70, 72; group, 66, 84, 110, 112; individual, 84, 110; qualitative, 82, 84, 111–112, 129, 166; quantitative, 5, 18, 67, 130, 166
database, 7–8, 34
Database of Scientific Materials, 8
Descriptive data, 68, 70, 72
diachronic, 55–56, 61, 63, 66
dissatisfaction, 70, 72, 82, 163
dissertation, 89, 95, 102, 104, 117–120, 124–125, 132, 159

189

dissertation supervisor, 117, 120, 124, 132
division of labor, 39–40
doctoral committee, 117
doctoral supervisor, 118
D–score approach, 61

Ebola virus, 10
editing service, 13, 154–156, 159, 169–172
editor, 10, 41, 45, 89, 95–96, 99, 107–108, 155–156
educational intervention, 81–82, 121, 125, 129, 131, 133–134, 137, 149, 156, 159, 161, 172
empirical, 3, 6, 13, 20, 26, 41, 47, 160
Englander, Karen, 3, 4, 9, 11, 42, 44, 48–49, 50–51, 121, 139, 165, 169, 171
English as a foreign language (EFL), 90, 167
English as a Second Language (ESL), 67, 70–71, 73–77, 79–82, 134, 139
English for Academic Purposes (EAP), 165
English journals, 9, 55, 60, 102, 154, 156, 173
English language publications, 9, 11, 13, 21, 35–37, 48, 110, 160–161, 167
English literacy, 4–6, 9, 115, 123
English writing, 12, 47, 49–50, 73–75, 81, 85, 93, 95, 104, 106, 109–110, 127, 129–132, 156
Ensenada, Mexico, 58
epistemic, 29
European Union, 9, 19, 49
Expanding Circle countries, 33
expert–insider, 22, 26, 29
explicit language courses, 126, 129
explicit teaching, 140–141, 143, 171

Faculty Writing Circle, 149–151, 153–154, 159, 172

feedback, 102, 114, 116, 118–119, 123, 127, 149–150, 152, 159
financial support, 13, 167
first author, 84, 88, 90, 94, 97, 100, 104
first language speakers (L1), 34, 53, 56, 61, 68, 70, 72–73, 76, 79, 81, 114, 127–128
Flowerdew, John, 3–4, 11–12, 22, 37, 42, 47–49, 51, 164–165
Flowerdew, John, 3–4, 11–12, 22, 37, 42–43, 47–49, 51, 164–165

Franck, Gregory, 18
functional linguistics, 50

Garfield, Eugene, 16
generalizability, 20
genre, 13–15, 29–30, 32, 50, 74, 80, 114, 117, 120, 145, 150, 154–155, 165–166, 168
geophysics, 9, 59
GeoRef, 7
grant writing, 91
grants, 91, 115, 144, 162, 169
Gross, Alan, 6, 14, 21–24, 26–28, 30
group data, 66, 84, 110, 112

Hanauer, David, 3–4, 11, 14, 114, 189
Harris, Randy, 15
hedging, 29, 48, 53
Hotellings, Harold, 72
Hyland, Ken, 15, 29–30, 32, 51–53, 165

Individual data, 84, 110
inferential statistics, 73
informational complexity, 21–22
Inner Circle countries, 33
institutional support, 117, 147, 160
interdisciplinarity, 38
International English Language Testing System (IELTS), 106

international: exchange programs, 122; journals, 8, 10, 36–37, 58, 86, 89, 97, 151, 169–170; publications, 36–37, 53, 175; scientists, 37, 53, 160–161, 164–165, 171, 173–174
intertextual, 17, 23, 28
Introduction-Methods-Results-Discussion (IMRAD), 6, 25; Conclusion, 27, 96; Discussion, 6, 27, 65, 95; Introduction, 6, 26–27, 32, 55, 67, 83, 95, 111, 160; Methods, 6, 26, 47; Results, 6, 27, 68, 131

jigsaw writing, 48
journals, 6–7, 11, 18, 20, 34–35, 40, 43–46, 53–54, 56–58, 60, 87, 96, 99–100, 107, 120, 123, 143–144, 149, 152, 154, 156, 165, 172–173; English, 9, 55, 60, 102, 154, 156, 173; international, 8, 10, 36–37, 58, 86, 89, 97, 151, 169–170; medical, 10, 41, 173; national, 34; non-English, 10–11, 36; peer-reviewed, 7, 34, 90, 94, 97, 173; regional, 10–11, 169; scientific, 6–9, 17, 35, 45; social science, 35
junior faculty, 63–64, 66–67, 76–81, 83, 97, 103–104, 109, 111–112, 120–121, 123–124, 130, 134, 148, 163, 175
junior scientists, 84, 97, 103, 111, 120–121, 123–124, 130, 164

Kaplan, Robert, 4, 53, 174
Kuhn, Thomas, 15

L2 science writing, 5, 11, 56, 61, 65, 67–68, 70, 76, 78–79, 83, 114, 127, 134, 136–137, 139–140, 160, 166, 175
labor, division of, 39–40
laboratory reports, 87, 95, 108, 114
Lancet, 10
language brokers, 21, 38, 50

language writing burden, 68
Latour, Bruno, 15, 18
learning events, 111–113, 121–122, 167
Lillis, Theresa, 7, 21, 33, 35, 37–38, 40, 42, 47, 50, 119, 139, 155, 165, 169–170
linguistics: applied, 14, 32, 40, 42–44, 140; corpus, 51, 53, 128; functional, 50
literacy, 3–5, 14, 16, 21, 55, 64–65, 114, 116–117, 119, 121, 123–125, 130–132, 135–138, 155, 158–159, 166
literacy, English, 4–6, 9, 115, 123

magazine, 108, 151
MANOVA, 72
Matsuda, Paul Kei, 11, 42–44
mean, 24, 69–71, 77–78, 87, 163
medical journals, 10, 41, 173
mentoring, 47, 98, 124, 147–149, 157–158, 172–173
merit pay, 8–9, 58, 108
metaphor, 37
method, 6, 26, 28, 30, 44, 61, 63, 88, 91, 109, 137, 141, 145, 148, 150–151, 155
Mexico City, 59
mixed method approach, 56
modal forms, 102
monolingual, 6
mother tongue, 49–50, 87, 109
multilingual, 12, 41, 114, 129, 164, 173
multimodal, 4

narratives, 26
national journals, 34
National Organization of Researchers, 60, 108
natural sciences, 41, 59, 63, 89, 128
non-Anglophone, 7, 35, 37, 40, 46, 160, 165–167

non-English journals, 10–11, 36
Non-Native Speakers of English (NNS), 4, 143
noun phrase, 22–23, 25, 29, 31
novice-outsider, 29

Organization for Economic Cooperation and Development (OECD), 8–9, 19, 34, 36, 58, 171
Outer Circle countries, 33
outline, 48, 55, 135, 154, 160

parochialism, 43
participant-scientists, 114, 116, 118–119, 126
passive voice, 23–24
pedagogy, 5, 43, 84, 111, 121, 126, 129, 131, 135, 140, 143, 165
peer-reviewed journals, 7, 34, 90, 94, 97, 173
persuasion, 15, 29
popular press writing, 126
postdoctoral fellowship, 47, 97–100, 103
post-method approaches, 135
pragmatic approach, 164–165
profiles, 84, 89, 97, 103
publication, 8, 10–11, 15, 17, 19–20, 38, 40, 43–44, 47, 49–51, 56, 58, 61–62, 64–65, 73, 81, 84–85, 90, 96, 100, 105, 107, 110, 115, 118, 120, 124, 126, 129–130, 134, 146, 149, 157–159, 171, 173, 175; English language, 9, 11, 13, 21, 35–37, 48, 110, 160–161, 167; international, 36–37, 53, 175; language of, 7, 51; scientific, 4–6, 9, 33–34, 36, 45, 50, 65, 67–68, 80–81, 85–87, 115, 117, 129, 133, 138–139, 146, 167; value of, 18, 31
publishing, 4, 6, 7, 11, 15, 17–18, 32, 35–36, 40–41, 46, 54–56, 58, 60, 64, 73, 87–89, 97–98, 100, 105, 107–109, 114, 116, 140, 152, 160, 164, 167, 173; in English, 3–4, 8, 10, 13, 36–37, 53, 58, 66, 134, 160–161, 163–164, 166, 169–171; international, 36–37, 40, 53, 166, 141, 170–171; multilingual, 12; non-English, 35–36, 160, 165, 169; process of, 150; scientific, 3–4, 19, 45, 57, 60, 73–74, 134, 146, 157, 166
Pysch Info, 7

qualitative, 5, 13, 36, 56, 61, 63, 66, 82–84, 110–112, 115–117, 126, 129, 131, 166
quantification, 30, 67, 80
quantitative, 5, 13, 18, 56, 61, 63, 66–68, 82–83, 126, 130, 166
quasi-experimental design, 70

reader-responsible text, 52
reading across the disciplines, 169
regional journals, 10–11, 169
research and development (R&D), 34–36, 39, 59
research article: characteristics, 20–21, 25–28, 30–31, 87, 120, 137, 139, 141; citation, 16–17, 53; genre, 6, 13, 15–18, 20–21, 25–26, 28–31, 51, 96, 112, 115, 117–118, 123, 137, 159, 166; history, 21–22, 29; medical, 36; producing, 19, 21, 140, 159; publishing, 15, 17–18, 20, 31, 110, 115, 119–120, 139, 158–159, 164; scientific, 14–15, 22–23, 25, 30, 32, 51, 80, 83, 85, 136–139, 158; second language writing, 32, 61, 64, 67, 75, 80, 82, 84, 111–116, 121–122, 124, 126, 129, 134, 136–137, 140, 156, 158; social function of, 19, 29; standardization, 21–22, 25; writing, 19–21, 23, 47, 56, 61, 64, 67, 75, 80, 82–84, 86, 111, 119, 125–128,

132–138, 144, 157, 159, 163, 166–167
research center, 4, 12, 59
research institute, 59–61, 63–64, 66–67, 70–71, 73, 75–76, 78–81, 84–85, 88, 97, 102, 109, 111–112, 121–124, 126, 130–131, 134, 162–163, 175
research writing, 19–20, 23, 61, 75, 80, 82–83, 86, 111–112, 114, 116, 118–122, 126–129, 133–139, 157, 163, 166
results, 6, 13, 25, 27–28, 40, 43, 48, 55–56, 63, 66–68, 70, 72–73, 76–77, 79–83, 85–86, 95, 98–99, 102, 107, 109, 130–131, 133, 151, 167
reviewer, 40, 43–44, 91, 100, 103, 120, 123, 127–128, 145
rhetoric, 6, 14–15, 26, 28, 38, 47, 51, 53, 140, 143, 145, 150, 154–156, 165, 168, 172
rhetoric and composition, 140

Salager-Meyer, Francoise, 7, 10–11, 15, 24, 29–30, 32–33, 35, 37, 41, 45, 51–52, 131, 169, 173
Science, 7, 9, 11, 19, 34–35, 45, 56, 58, 60, 138, 141, 143, 147, 161, 173
Science Citation Index, 34, 60
scientific: journals, 6–9, 17, 35, 45; notebooks, 114; publications, 4–6, 9, 33–34, 36, 45, 50, 65, 67–68, 80–81, 85–87, 115, 117, 129, 133, 138–139, 146, 167; reports, 114; writing, 4, 5, 11–15, 22–25, 46, 50, 52–56, 61–63, 66–68, 70–71, 73, 76–83, 91–92, 94–95, 97, 106, 108–111, 113–118, 120–125, 127–128, 131–133, 140, 144, 146, 149–150, 154, 157–158, 161–162, 166–168
Scientist, The 45

scientists: international, 37, 53, 160–161, 164–165, 171, 173–174; junior, 84, 111, 120–121, 123–124, 130, 164; participant scientists, 114, 116, 118–119, 126; second language, 3–5, 11, 13, 32, 48, 56, 61, 67–68, 70, 74–76, 80, 134, 143, 160; senior, 66, 84, 114, 121–124, 130, 134, 138, 164; student scientists, 168
second author, 84, 86–90, 97
second language: editing, 154–156; learning English as, 100; literacy, 4–5, 112–113, 115–116, 123–125, 174; scientists, 3–5, 11–13, 32, 48, 55, 61, 67–68, 70, 73–76, 80, 83, 120, 122, 126, 128, 134, 143, 155–156, 159–160, 162, publishing in, 4–5, 11, 13, 50, 61, 67–68, 134, 137, 139, 170; ratings, 81; scientific writing, 5, 11, 56, 61, 65, 67–68, 70, 76, 78–79, 83, 114, 127, 134, 136–137, 139–140, 160, 166, 175; teaching of, 11, 56, 137, 156–158; writing, 5, 11–12, 32, 46–47, 53, 55–56, 67–70, 73–76, 79–81, 97, 110–112, 115–117, 119, 125–126, 129, 131–132, 135–136, 140, 159, 164, 166, 174
second language speakers (L2), 5, 56, 61, 65, 67–68, 70, 76, 78–79, 83, 114, 175
senior faculty, 47, 63, 66–67, 76–81, 83–85, 89, 97, 103, 109, 111–112, 114, 116, 120–125, 128, 130, 134, 138, 148–149, 164, 172–173, 175
senior scientists, 66, 84, 89, 114, 121–124, 130, 134, 138, 164
Sistema Nacional de Investigadores (SNI), 10
social science journals, 35
social sciences, 39, 58, 60
Society of Exploration Geophysicists, 9

sociology of scientific knowledge, 50, 140
standard deviation, 68–71, 76–78
student advising, 60, 105
student-scientists, 168
study abroad, 114
style editor, 96
subheadings, 25
subject-verb agreement, 102
support: administrative, 139; financial, 13, 167; institutional, 117, 147, 160
survey, 41, 61–64, 66–69, 71, 73–75, 80–82, 121
Swales, John, 4, 6, 25–27, 37, 43, 47, 50–52, 154, 165
synchronic, 55–56, 61, 66
syntax, 45, 47, 101, 150, 168

teaching, 10–11, 13, 59–60, 63, 67, 70–71, 73, 75–77, 79–81, 89, 104–105, 109, 132, 134, 140–143, 148, 162–163, 171–172
teaching university, 63, 67, 70–71, 73, 75–77, 79–81, 109, 134, 163
technical writing, 101, 103, 108, 126
TESOL Quarterly, 44
Test of English as a Foreign Language (TOEFL), 36, 90, 104
Texas A&M University, 143
The Scientist, 45
thesis, 90, 95–96, 99, 104, 116–117, 124–125, 132, 158
thesis supervisor, 116–117
Thomson Reuters Web of Science, 7–8, 34–35, 53, 56
translation, 13, 48, 76, 87, 92-93, 95-96, 101-105, 107, 110, 128, 154-156, 159, 163, 169-173
translation service, 93, 105, 110, 128, 154-155, 163, 172
translator, 48, 50, 90, 105, 155-156
tutor, 128-130, 132, 145

Universidad Autónoma de Baja California (UABC), 58
user-friendly, 22
Uzuner-Smith, Sedef,9, 37, 41, 47-48, 169, 171

Woolgar, Steve, 15, 18
World Bank, The, 8, 10, 19, 58
writer-responsible text, 51-52
writing: academic, 11, 47, 50, 53, 88, 101, 117, 143, 145, 164; collaborative, 98, 100, 102, 123-124, 133; English, 12, 47, 49-50, 73-75, 81, 85, 93, 95, 104, 106, 109-110, 127, 129-132, 156; grant, 91; jigsaw, 48; L2 science writing, 5, 11, 56, 61, 65, 67-68, 70, 76, 78-79, 83, 114, 127, 134, 136-137, 139-140, 160, 166, 175; popular press, 126; process, 20-21, 24, 47, 56, 63, 65-66, 74, 80, 86, 99, 116, 119, 128, 132, 138, 159; research, 19-20, 23, 61, 75, 80, 82-83, 86, 111-112, 114, 116, 118-122, 126-129, 133-139, 157, 163, 166; scientific, 4-5, 11-15, 22-25, 46, 50, 52-56, 61-63, 66-68, 70-71, 73, 76-83, 91-92, 94-95, 97, 106, 108-111, 113-118, 120-125, 127-128, 131-133, 140, 144, 146, 149-150, 154, 157-158, 161-162, 166-168; second language, 5, 11-12, 32, 46-47, 53, 55-56, 67-70, 73-76, 79-81, 97, 110-112, 115-117, 119, 125-126, 129, 131-132, 135-136, 140, 159, 164, 166, 174; technical writing, 101, 103, 108, 126;
writing across the disciplines, 169
writing center, 13, 116, 127, 129, 132, 138, 145
writing courses, 104, 139, 143, 158
writing process, 20-21, 24, 47, 63, 65, 74, 80, 82, 119, 128, 132, 138

Zerbe, Michael, 6, 15

About the Authors

David Ian Hanauer is Professor of English/Applied Linguistics at Indiana University of Pennsylvania and an educational researcher and the Assessment Coordinator of the Phage Hunters Integrating Research and Education (PHIRE) Program situated in the Hatful Laboratory at the University of Pittsburgh. His science education research has addressed assessment in the sciences, the processes and teaching of scientific inquiry and scientific writing in first and second languages. His literacy research has focused on the connections among authentic literacies and social functions in first and second languages and has addressed the genre specific aspects of poetry reading and writing in L1 and L2, graffiti, and linguistic landscape research. He is the author of six books including *Scientific Discourse: Multiliteracy in the Classroom, Poetry as Research* and *Active Assessment: Assessing Scientific Inquiry* (co-authored with Graham Hatfull and Deborah Jacobs-Sera). His articles have been published in *Science* and a wide range of applied linguistics and educational journals. He has received funding from the National Science Foundation, Howard Hughes Medical Institute and the U.S. Department of Education. He is co-editor of the *Language Studies, Science and Engineering* book series with John Benjamins.

Karen Englander, York University, Canada, is a long-time faculty member of at the Universidad Autónoma de Baja California, Mexico, where she works with scientists and graduate students who seek to publish their research in English. She has published empirical research on the policy, linguistic, and identity issues implicated in writing and publishing scholarly work in English when the writer is not a native speaker of the language. She is co-editor of *Discourses and Identities in Contexts of Educational Change*, and her work has appeared in the *Journal of Applied Linguistics, Discourse Studies, Journal of Language, Identity and Education, and Journal of International Women's Studies,* and *Written Communication*, among others.

www.ingramcontent.com/pod-product-compliance
Lightning Source LLC
Chambersburg PA
CBHW021214240426
43672CB00026B/176